THEORY OF USER ENGINEERING

THEORY OF **USER ENGINEERING**

Masaaki Kurosu

CRC Press
Taylor & Francis Group
Boca Raton London New York

CRC Press is an imprint of the
Taylor & Francis Group, an **informa** business

CRC Press
Taylor & Francis Group
6000 Broken Sound Parkway NW, Suite 300
Boca Raton, FL 33487-2742

© 2017 by Taylor & Francis Group, LLC
CRC Press is an imprint of Taylor & Francis Group, an Informa business

No claim to original U.S. Government works

Printed on acid-free paper
Version Date: 20160801

International Standard Book Number-13: 978-1-4822-3902-7 (Hardback)

Library of Congress Cataloging-in-Publication Data

Names: Kurosu, Masaaki, 1948- author.
Title: Theory of user engineering / Masaaki Kurosu.
Description: Boca Raton : Taylor & Francis, CRC Press, 2017. | Includes
bibliographical references and index.
Identifiers: LCCN 2016032447| ISBN 9781482239027 (hardback : alk. paper) |
ISBN 9781482239034 (ebook)
Subjects: LCSH: Human engineering. | User-centered system design. | Product
design.
Classification: LCC T59.7 .K87 2017 | DDC 620.8/201--dc23
LC record available at https://lccn.loc.gov/2016032447

Visit the Taylor & Francis Web site at
http://www.taylorandfrancis.com

and the CRC Press Web site at
http://www.crcpress.com

Printed and bound in the United States of America by Edwards Brothers Malloy on sustainably sourced paper

Contents

Preface

Until the 1990s, usability engineering played a key role in improving the design quality of artifacts and the quality of life (QOL) of users. However, when the concept of user experience (UX) became popular around 2000, the situation drastically changed. It widened the focus of stakeholders from the objective characteristics to the subjective impression of people who use them. It changed the concern from just the product to various kinds of artifacts including the service activity. It widened the temporal focus of QOL from the moment of usage to the whole length of temporal sequence including expectations and the continuous use afterward.

There are many definitions of UX, including those by Garrett (2000), Morville (2004), and Revang (2007). About 30 definitions are listed on the All About UX website (http://www.allaboutux.org/ux-definitions). Even the publication of "UX White Paper" in 2011 did not solve this chaotic situation, though it did give clear understanding of how UX is unique compared to the usability.

But what is more important than the concept of usability or UX is the consideration of users. Both usability and UX should be considered for improving QOL among users. It is important to understand that the goal of user engineering concerning usability and UX should not be toward increased sales. Indeed, marketing people are struggling to increase positive UX. But this is the point where the marketing approach and user engineering are different (Kurosu 2006). I should clarify that my position is on the side of users and the goal is not money but the QOL of users. This is the reason why this book's title includes the term *user engineering*.

This book consists of twelve chapters: The first chapter is about the concept of user and user-centered design (UCD), the second chapter concerns the relationship between usability and UX, and the third chapter is on the artifact evolution theory (AET). Chapters 4 to 10 discuss the design process and the business process. Chapter 11 is on the relationship of user engineering and recent technology, and Chapter 12 is about the future of user engineering.

I encourage readers to come into the field of user engineering. In some parts, user engineering looks like a combination of existing concepts. Indeed, many methods described in this book have already been developed by usability engineering and UX design. But because of the stance of this book, the reader should take care to catch the meaningful variation contained in those concepts and methods. We have to step forward to achieve and establish better experience among users.

Note: The use of the term *artifact* in this book means a human-made thing in general and includes the product (hardware and software) and the service (humanware). The words *artifact* and *product/service* are used interchangeably depending on the context.

Masaaki Kurosu

1 What Is User Engineering?

User engineering is a set of concepts and methods for improving the quality of life of users. Many concepts and methods have been inherited from usability engineering and user experience (UX) design. But the difference with user engineering lies in its stance:

1. Compared to usability engineering, its scope is wider and includes other quality characteristics such as reliability, safety, and even beauty and cuteness.
2. Compared to UX design that is currently a bit more oriented toward marketing and designing an attractive artifact for consumers to buy, it focuses on user satisfaction and the significance of artifacts.

1.1 WHO ARE THE USERS?

Simply speaking, users are those who use the artifact. But a bit more explanation is necessary in its relation to consumers. The marketing approach targets consumers and puts emphasis on figuring out their motivation to purchase the artifact. Generally, it is the goal of marketing that consumers should buy the artifact. Of course, marketing emphasizes the feedback for the next purchase and in this sense puts emphasis on the use of the artifact. That is, the marketing approach also looks after the user's response. But its main point is on the purchase of the artifact.

Consumers who purchase an artifact will become users. Typically, they start to use it by trial-and-error or by reading the operator's manual if there is one. Then users get used to the artifact and learn more about it and gradually grow from novice users to experts. After long-term usage, some event will occur in a positive sense or a negative sense. And when the performance of the artifact degrades or there is any malfunctioning, users will stop using it.

Hence, all through the artifact lifecycle, that is, before, during, and after the purchase, consumers change into users and obtain various kinds of information about the artifact and use the artifact itself. But, user engineering mainly deals with the phase of the user. It also deals with the phase of consumer because it is also the phase of preuser.

1.1.1 ISO/IEC 25010:2011

When users start using an artifact, ISO/IEC 25010 proposes a list regarding the concept of users. Figure 1.1 shows a revised version of the definition of *users* in the standard. At the top level, direct users who directly interact with the system are different from indirect users who do not interact with the system but use the output of

Direct user	Those who interact with the system	Primary user	Those who interact with the system for the achievement of primary goal	Active primary user	Those who interact with the system actively	MRI operator
				Passive primary user	Those who interact with the system passively	Patient
		Secondary user	Those who provide the support to the system			Maintenance personnel and manager
Indirect user	Those who don't interact with the system but use the output					Medical doctors who receive the test result

FIGURE 1.1 Classification of users. (Adapted from ISO/IEC 25010.)

the system. Take for example the case of an MRI at a hospital; the medical doctors who receive the resulting image are indirect users.

Direct users are, then, classified as the primary users who interact with the system for the achievement of a primary goal, or the secondary users who provide the support to the system. In the case of an MRI, the maintenance personnel and manager of the inspection department are examples of secondary users.

The classification of ISO/IEC 25010 is given at these two layers. But if we consider the case of an MRI, there are MRI operators and patients as primary users. So I added a third layer by distinguishing active primary users, who actively interact with the system, and passive primary users, who passively interact with the system. In the case of an MRI, operators are active primary users and patients are passive primary users.

But passive primary users do not always exist. If we consider the projector used in a classroom, active primary users are teachers who use the projector for teaching the lesson by showing PowerPoint slides to students. But students are not passive primary users but are indirect users who look at the slide images on the screen and get information from them.

For active primary users, the objective quality characteristics, including the usability (especially the ease of operation and the ease of understanding), are quite important. But for passive primary users, the exterior design of the MRI, for example, should give subjective quality characteristics such as the sense of relief for the comfort during the examination.

For secondary users, the cost, ease of maintenance, and reliability are important, and for indirect users, the ease of understanding the output is important. As such, different aspects of quality characteristics are important for different types of users. Generally speaking, there is a tendency to use the word *user* for meaning the active primary user. But at the same time, it must be remembered that there are other types of users and the artifact should be designed considering all types of users as much as possible.

In the example of the projector used in the classroom, the (active) primary user is, of course, the teacher who operates the projector. The secondary user is the personnel of the school who do the maintenance work, for example, replacing the bulb in the projector. Indirect users, in this case, are students who watch the projected slide images and ask questions on the information contained on the slides.

1.1.2 STAKEHOLDERS AND USERS

From Figure 1.1, one might suggest that users might be the same as stakeholders. But stakeholders of an artifact will include those people who belong to the industry including manufacturers and service providers as well as those who are outside the industry. Thus, user-centered design (UCD) is an outside-in approach, whereas marketing is an inside-out approach. User engineering focuses on the outside and takes an outside-in approach.

1.1.3 USERS AND CONSUMERS

Consumer is the term that is frequently used by marketing people. It is quite natural that they use this word because the word *marketing* comes from the word *market*, and the market is the place where money and the product are exchanged between the consumer and the merchant. But, it should be noted that the activity at the market will end when the exchange is completed. After the purchase, the consumer will go back to their ordinary life to become the user.

The end of marketing activity is the time when users start their activity. More precisely, former consumers will become users and start using the product they purchased. As a natural result, marketing people are not much interested in the life of users and problems that users may face, but are focused more on how to sell products to consumers. Hence, they are not much concerned with usability. This is because usability is difficult to evaluate at the marketplace. From this viewpoint, it is a bit strange that marketing people use the term *UX*. UX is the experience of users and marketing people are not concerned much about the life and the problems among users in their ordinary life. It is suspected that marketing people are using the term *UX* in the sense of "anticipated UX" that is not yet confirmed by the real experience among users. Though it is true that the anticipated UX can be a part of the total scope of the UX, the hit ratio of the anticipation is usually not high and we will have to survey the actual UX in the actual context of use.

1.1.4 INTENDED USERS

Sometimes, the term *intended user*, or *targeted user*, is used among developers. It is a common habit of marketing people to specify who the main user is. The persona method is frequently used for the purpose of sharing the image of intended users among stakeholders. This approach is effective for curtailing the development cost by focusing on the limited number of specific users so that the specification can be easily decided.

But this is a tricky strategy. Strictly limiting the range of users only to the intended users will result in the loss of chance for the artifact to be used by a wide range of people. Hence, the adequate strategy is to think of the loose bell-shaped user population where the center of the distribution is the targeted user. This strategy will lead to a compromise with the notion of universal design that will be discussed in the next section.

1.2 UNIVERSAL DESIGN AND ACCESSIBILITY

Until the 1990s, it was common to implicitly assume that the user of any product or system is a male in his 30s with a certain level of technological skill and intelligence. But it actually was the image of the engineers themselves. As a natural result, there occurred the discrepancy between the actual user and the assumed user image. Furthermore, there were complaints among users that the artifact they purchased did not fit them. Many of the real users are not males in their 30s with a certain level of skill and intelligence.

What changed this situation was the advent of the concept of universal design (Lidwell et al. 1997). Universal design is defined by Mace (2016) as "the design of products and environments to be usable by all people, to the greatest extent possible, without the need for adaptation or specialized design." He also wrote, "The intent of universal design is to simplify life for everyone by making products, communications, and the built environment more usable by as many people as possible at little or no extra cost. Universal design benefits people of all ages and abilities."

In this sense, universal design aims at the whole variety of users, including sex, age, disability, culture, and language. A possible list of such characteristics will be explained in the next section. But because many of those who started to work in the field of universal design also worked in the field of accessibility, the activity of universal design frequently puts emphasis on disabled and senior people rather than the whole range of diversity. According to the definition in ISO/IEC 25010, *accessibility* is "to what extent does the system need to be effective, efficient, risk free and satisfying to use for people with disabilities." In this sense, accessibility is the usability for the disabled (refer to Chapter 2 for the definition of *usability*).

User engineering focuses on the possible widest range of people in the same sense of genuine universal design.

1.3 DIVERSITY OF USERS

There are three different aspects on the diversity of users: (1) characteristics, (2) inclinations, and (3) situation and environment (Figure 1.2).

1.3.1 CHARACTERISTICS

Characteristics belong to users themselves and represent the differences among users. They include such physical differences as age, sex and gender, disability, temporary situation, physical traits, language, knowledge and skill.

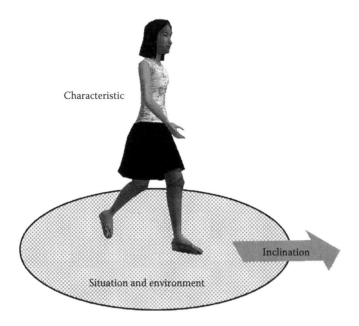

FIGURE 1.2 Three aspects of diversity among users.

1.3.1.1   Age

Age is a characteristic along the time dimension. There are three different aspects for a person in terms of the time dimension—age group, cohort, and same time period—as shown in Figure 1.3.

Age groups are those of the same age. An example is the definition of the senior person, those equal to or older than 65 years, according to the World Health Organization (WHO). Physical strength, physiological functioning, and psychological ability will show a decreasing tendency, especially after 60 to 65 years of age. This tendency is almost the same as it was in the 1900s and 2000s and maybe later in the 2050s, though there have been some positive changes due to improvements in living environment and medical science. Regarding the use of ICT (information and communications technology) products and systems, people show a similar set of changes as they grow older: presbyopia makes them need to see larger letters and use reading glasses, the loss of hearing ability makes them need louder auditory signals and a clear pronunciation or a hearing aid, slower reaction time requires longer waiting time, lower visual search ability requires a simple visual interface, and so on. On the contrary, children will have to be given special care on the user interface: for example, messages from the system should use simple and plain linguistic expressions so that they can understand the meaning.

A cohort is a group of people who were born during the same era and experience similar social and historical events. A generation, such as the "digital native," is an example of a cohort. Cohorts who were born before the time of ICT development generally show poor literacy for using such devices and systems.

Finally, all people are living in the same period of time. Those who are living in the 2010s, whatever their age, are provided the chance of using the smartphone,

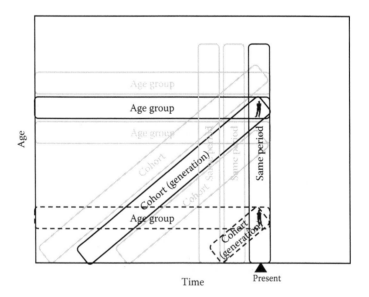

FIGURE 1.3   Three temporal aspects of a person.

a PC, and the Internet. And we'll have to be careful to note the difference of age groups and cohorts in terms of background skill and knowledge.

1.3.1.2 Sex and Gender

Almost half of people are male and the other half are female. Sex is a biological concept and gender is a sociological concept.

Regarding sex, there are physical and physiological differences that should be considered when designing a user interface. For example, males on average have stronger muscular strength and larger hands than females. This is related to the optimal weight of portable devices and the optimal size of keyboards. Similar issues can be found on the key pitch of a piano—it is usually too wide for those with small hands.

Physical differences are also related to sports. With the exception of horse riding and some other sports, sporting events are usually held separately for men and women competitors. This is accepted by most people and not regarded as gender discrimination.

Gender is strongly influenced by culture and history. Especially *Kansei*, or the sensibility aspects of people, shows a big difference between men and women. It has been believed in many societies that men generally tend to pursue power and logic, while women tend to appreciate beauty and emotion. But this tendency might be the result of culture and education. For example, the single-lens reflex digital camera is black in color based on the idea that many users of this camera are men. But this might be a cultural bias. And when laptops and smartphones started to be used by many women, colors were extended to include champagne gold, pink, red, white, and so forth. The same is true for the square design for men and the round design for women. But it is true that many women (at least in the Japanese market) prefer the round design in gold or pink. It is unknown whether these tendencies will survive in the same way in the future. It will be in line with the change of social attitude of people toward gender.

An interesting phenomenon is the existence of the "women-only car" in the railway system in Japan (Figure 1.4) and in other countries, including Iran, Pakistan, and India. The purpose of the women-only car is twofold: one is religious and the other is protection of the weak. In Japan, the women-only car was introduced around 2000 for the purpose of protecting women from sexual harassment in a crowded car.

Problems related to the gender cannot easily be solved. The mind-set of people or the social system will change the situation.

1.3.1.3 Disability

Previously, disability was simply considered a lack or the decline of physical or mental functioning. And those who have a disability were regarded as handicapped. However, in 2001, the International Classification of Functioning, Disability and Health (ICF) was endorsed by the World Health Organization's World Health Assembly (WHA) and today the ICF is the worldwide framework for health and disability.

In the ICF model, there are two groups of factors. First group is a fundamental factors including "body functions and structure," "activity" and "participation" are

FIGURE 1.4 Women-only car (Japan).

regarded as the first group influencing the health condition (disorder or disease). Body functions are physiological functions of body systems including psychological functions. The body structure is the anatomical part of the body, such as organs, limbs, and their components. These two factors are at the biological level. Activity is the execution of a task or action by an individual. This is at the individual level. Participation is the involvement in a life situation. This is at the social level.

The second group of factors is contextual factors, which include environmental and personal factors. Environmental factors make up the physical, social, and attitudinal environment in which people live and conduct their lives. Personal factors are related to personality, emotion, motivation, and preference of an individual.

Hence, the scope of ICF is not only at the biological level but also at the individual and social levels. And the concern of ICF is not only for diagnosing the disability but also for providing the improvement of quality of life (QOL) among those who have the disability with consideration for participation and environmental factors. Environmental factors for those with disabilities include the support by the development of new devices and systems, electric wheelchairs, and ICT-related supporting aids such as the screen reader.

One point that should be addressed here regarding disability is the ratio of people with color blindness. Because of the biological nature of chromosomes, the rate of color blindness is about 5% for men and even lower for women (in Japan). Designers of the visual interface should take great care when using color coding. Adequate labels should be added to the colored marks and lines. In addition, the use of light emitting diode (LED) that changes from green to red and vice versa is not adequate.

1.3.1.4 Temporary Situation

There are other types of disabilities or inconveniences that do not last forever. For example, pregnancy causes many changes for women including increased body weight, difficulty maintaining balance, morning sickness, difficulty bending forward, and difficulty standing for a long time. In support, for example, many train companies provide "priority seats" for pregnant women in addition to other people with difficulties. But many of the difficulties related to pregnancy will last for less than a year, thus pregnant women should be treated as having a temporary disability.

There are other inconveniences, for example, the fracture of a leg bone, which oblige people to use crutches for a short time. Priority seating on transportation systems for people with such inconveniences are needed. Furthermore, those with heavy luggage will have difficulty going up and down the corridor at the station or in other places. Escalators and elevators should be provided for them as well as for others who have difficulty moving along the corridor.

Various artifacts have been developed for supporting these people. But there are still many problems that should be solved. One example is the reading of prepaid transportation cards at the station. If people have a lot of luggage or both hands are occupied, they will have to stop at the ticket gate, put down the luggage, then take the card out of the pocket. If the technology of the Internet of Things (IoT) will be able to allow them to pass the gate without taking the card out of the pocket, the flow of people at the gate will become smoother.

1.3.1.5 Physical Traits

Statistical data on the individual differences of physical traits can be found in the anthropomorphic database. Ergonomics has accumulated an enormous amount of data for the design of various products and systems.

People differ in such static characteristics as length, weight, shape, and body color, as well as such dynamic characteristics as angle of rotation, length of expansion, and speed of motion. Hence, artifacts should be designed by adapting to the varieties of users. Customizing the artifact and zoning the people are two strategies to adapt to those differences. Customization involves installing an adjustable mechanism, such as the lever on a chair to adjust its height and the location of holes on a belt. Zoning means segregating people into several categories, such as how clothing sizes are classified as S, M, L, and XL.

Biometrics is another type of technology that can be used to identify the individual in terms of the pattern differences of the fingerprint, palm, face, iris, and so forth. This technology will provide the user simpler and easier ways to be identified instead of an ID card. People may not need to carry anything but their body to be identified.

1.3.1.6 Language

A major issue with user interface design for different ethnic groups is the use of language. There are thousands of languages in the world. Because language is a major tool for communication, each artifact should be customized with regard

FIGURE 1.5   Traffic sign in Japanese and Portuguese in Hamamatsu, Japan.

to the language of the user including the sign, label, and user's manual. But customization to thousands of languages is actually impossible, so the common language approach is usually taken. Quite often, English is used as the common language. But we should realize that there are lots of people who do not understand English. Hence, products and systems exported to South America should use Spanish (except Brazil where Portuguese is used). At the railway station in Tokyo, Japanese, English, Chinese, and Korean are used for the guidance display for providing adequate information to passengers. Because Japan is located in East Asia, it is natural for Japan to adopt the language of its neighbors—China and Korea—as well as English for public guidance. An interesting example is the traffic sign in the city of Hamamatsu (Figure 1.5). Because there are many Japanese-Brazilian people working at factories in Hamamatsu, this sign uses Portuguese, instead of English, and Japanese.

Another approach to overcome the issue of language is to use signs and symbols. An example is the symbol of toilet of which men are usually represented as a silhouette of a person with trousers and women represented as a person with a skirt. Sometimes language is added, usually English, but most people can understand what they mean even though there are many women with trousers today. Another example is to use video symbols instead of words for time-related media manipulation, such as the square for stop and the double bar for the pause.

In the near future, public signage will adopt IoT and users will be able to see the translated message on their cellphones. This technology will allow everybody in the world to get the adequate information in their mother tongue.

1.3.1.7 Knowledge and Skill

Not all users have a sufficient level of knowledge and skill to use an artifact. Providing the information at the level of beginners is important, but even among beginners, the level of knowledge and skills that should be presupposed are different depending on the type of artifact. For example, about 30 years ago, the level of knowledge and skill of novice users were very low in terms of the computer. But now, the cohort has shifted and there are those who are categorized as digital natives. And most users today have a mental model of how the computer works and have enough skill to use the basic functions of a computer.

Regarding knowledge and skill, there are novices and experts. The progress from novice to expert takes the form of the learning curve. Usually the curve is S-shaped, that is, at the beginning the curve performance does not go up much regardless of the effort, but after a certain period of time it will go up and finally reach the saturation level. The user interface should be designed considering the ease of understanding and the ease of operation, especially for novices. But expert users have a tendency to prefer efficiency rather than an easy but time-consuming procedure. The user interface design should take into account both types of users.

It should also be noted that novice users do not frequently read the user's manual before using the artifact. Although a good manual is important, the artifact itself should be designed to be easy to use so that it can be operated without the manual.

1.3.2 Inclination

Inclination is a dynamic aspect of the user and includes taste, innovativeness, religion, culture, and attitude.

1.3.2.1 Taste

The taste for artifacts differs from user to user. Taste may be influenced by individual factors such as sensitivity, personality, and personal experience, and by social factors such as life history, social environment, and natural environment. Kansei engineering and affective engineering focus on the issue of personal taste so that the product will attract the attention of a wide range of consumers. But the precise prediction is very difficult and all they can do is just explain the resulting fact.

The selection of color, shape, and material is much influenced by taste. For example, coloring strategy of the product deliberately examines the trend of color preference, the psychological image of color, and other factors. Many products appear in the market with several colors so that customers can make their own choice.

1.3.2.2 Innovativeness

According to Rogers (1963), consumers can be classified into five types: innovators, early adopters, early majority, late majority, and laggards in terms of the attitude toward new products. Innovators are sensitive to the novelty of artifacts and early adopters play the role of opinion leaders. The most conservative type is the laggard who sticks to the traditional way of life.

In Rogers's classification, innovators and early adopters usually attract the attention of marketing. Indeed, people have a general tendency of being attracted by the novelty of artifacts. And it is necessary for the industry to diffuse new products and new services into the market. But what we should focus on from the viewpoint of user engineering is the life of laggards. They are conservative and adhere to artifacts that they have been using. See Chapter 12, Section 12.2 regarding the life of laggards in relation to sustainable society.

1.3.2.3 Religion

Religion influences the way of life of people. Sometimes, ICT can support people who believe in a specific religion. An example is the Prayer Compass by Casio, which is a digital watch with the function to show the direction of Mecca and the time to pray. In the normal situation, the mosque is built in the direction of Mecca. But for traveling Muslims, Casio's digital watch, which is equipped with GPS and a compass, calculates the user's current location and displays the direction of Mecca and the time to pray. This model is said to have sold more than 400,000.

The use of the right hand for eating in the Islamic world is mentioned in Chapter 3, Section 3.2.8. In this case, the artifact is not the product but the part of body. This example also shows that sometimes religion plays a key role in the use of artifacts.

1.3.2.4 Culture

Usually, when we talk about culture we talk about the national culture (Hofstede 1991). Indeed, the circulation of people and products is restricted by the constraints imposed by the nation. And people in each nation have created their own culture. That is the reason why the national culture is taken up, even in academia, as a synonym of the generic term *culture*.

But there are various types of culture, as shown in Figure 1.6. This onion model of culture shows that there are individual culture, family culture, district culture, country culture, continent culture, and global culture. Individual culture is specific to an individual and it could be called the behavioral pattern of a person or the way

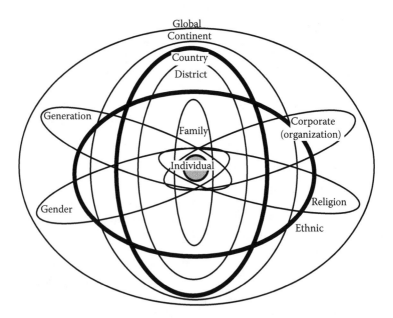

FIGURE 1.6 Onion model of culture. (Adapted from M. Kurosu, 2005, "How Cultural Diversity Be Treated in the Interface Design? A Case Study of e-Learning System," HCI International 2005.)

of thinking of a person. The next level is the family culture. No family is the same depending on the individuals in the family, the natural and social environment of the family, and other factors. Family members who do not follow the family culture will be alienated. Similarly, at the next level there is a district culture. The district may take different levels from the individual town to regions of the country. For example, in the United States the city of Boston has its own culture as a town whereas the East Coast has a different culture from the West Coast. Western Japan (e.g., Osaka and Kyoto) is different from Eastern Japan (e.g., Tokyo). The national culture is the next level. The major difference between the country culture and the district culture is that it allows people and artifacts to move freely among districts within the same nation. People do not have to pay a tax to move from the eastern part of a country to the western part in most cases. But there are differences of pronunciation, wording, customs, and so on between them, although they fundamentally use the same language as the mother tongue, watch almost the same TV programs, celebrate the same holidays, and so on. Between the country culture and the global culture, there is a continent culture. There are many expressions for the continent culture: European culture, Asian culture, African culture, and so forth. For example, Japan, China, and Korea have something in common as the East Asian culture, but each of them has its own unique culture. The last level is the global culture where all the people in the world have something in common as human beings. One example is that all the people in the world have initiation ceremonies such as marriages and funerals.

Besides these culture layers, there are ethnic culture, generation culture, gender culture, corporate (organizational) culture, and religious culture. One important point is that those who have the same country culture may have a different ethnic culture. An example is Chinese people who live in various parts of the world within Chinatowns. A Chinese person living in the Chinatown of San Francisco may adapt to the American culture, while retaining parts of the Chinese culture.

Because culture dominates the behavioral pattern of people in its broadest sense, designing of an artifact should take the deepest care in terms of localization. This is the reason why the idea of universal design is important.

1.3.2.5 Attitude

Spranger, an early twentieth century German philosopher, proposed six different types of value attitudes. They are theoretical, economic, aesthetic, social, political, and religious. Although this typology applies to the way of living and cannot directly be applied to the attitude of buying and using the artifact, there are different attitudes and value systems among consumers and users. Some people weigh much on the price, whereas others weigh the functionality, the beauty of the design, or the novelty.

An interesting point is that those who put emphasis on the price do not always take the same attitude in every case. Their attitude may change depending on the type of artifact. Some purchase inexpensive appliances and small houses, but they may pay more for accessories if they put much emphasis on the aesthetic. Some pay a lot of money for a new model of the smartphone even though they are living in a small room and eating junk food.

1.3.3 SITUATION AND ENVIRONMENT

Even within the same person, the behavior pattern may be different from ordinary ones depending on situational and environmental factors.

1.3.3.1 Emotional State

Usually, people conduct their behavior under the normal emotional state (i.e., not in a strong rage nor in deep grief) and usually in a mild and positive mood. Designers assume that their products will be used under a normal emotional state. As a result, most products are designed to provide neutral emotional value. The refrigerator, washing machine, vacuum cleaner, cash register, and ballpoint pen are such examples.

But some products and services are designed to give a certain level of positive feeling. This is where Kansei engineering works. Barbie dolls, stuffed bears, and Hello Kitty are such products, and a Japanese airline company painted Pikachu characters on the body of airplanes. They are expected to evoke the positive feelings of love, cuteness, and attachment. Treating customers in a pleasant manner is an example of service to evoke a positive feeling.

Although Kansei engineering is seeking to produce something that brings a positive feeling, human beings are not always seeking artifacts that trigger a positive feeling. Sometimes people read sad stories or watch tragic movies intentionally and move themselves to tears. It is human nature to seek catharsis.

On the other hand, people use ordinary goods in an unusual way when they have a strong feeling. For example, ceramic plates are made for dining. But when couples fight, plates can become artifacts that are thrown and broken. A bronze statue may be too heavy and cause serious harm and beautiful Baccarat glassware is too expensive to be broken. Thus, ceramic plates will be chosen as the artifact to express the emotion of anger.

1.3.3.2 Arousal Level

Arousal level is known to influence the level of performance. If it is at a low level, that is, people are asleep or drowsy, the performance level is low. They cannot complete the complex task and it will take a long time to achieve the goal. Usually, people are at the optimal range and perform every activity adequately. But in a hazardous situation such as an accident or fire, level of performance decreases in spite of the increased level of the arousal level (Figure 1.7).

Design of artifacts should consider the arousal level they will be used at. An example at the low arousal level is the alarm clock. It wakes people from the sleeping state when the performance level is almost zero. Usually, the alarm clock is equipped with a stop button on top of its body that can be pushed or touched by the very simple action of hitting it.

A new technology is being developed to detect the arousal level of car drivers by their body movement, eye size, rotation angle of the steering wheel, the distance change between the car ahead, and deviation of the car from the center line. Such technology will help prevent possible accidents caused by the low performance due to a lower arousal level.

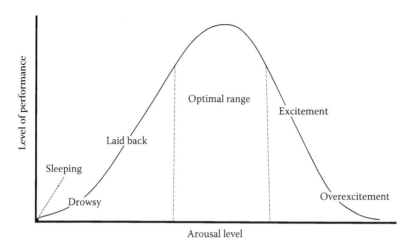

FIGURE 1.7 Arousal level and performance.

An example of an artifact that will be used at high level of arousal is the life jacket on an airplane. The situation when passengers are requested to wear a life jacket is a serious one and most people will be at an extreme level of arousal and some of them will lose control of their normal behavior. This is why the video on how to wear the life jacket is shown repeatedly before the departure.

Rasmussen (1986) proposed a distinction among knowledge-based behavior, rule-based behavior, and skill-based behavior. Skill-based behavior is instantly triggered by the perception of stimulus and will take a short time to be completed. Rule-based behavior is triggered based on the rule stored in long-term memory and will take a bit longer to recall than skill-based behavior. Knowledge-based behavior is done based on the thought process and the decision process in the cognitive system. Hence, it will take a long time until the behavior is started.

The emergency on the airplane is a situation that requires the identification of what is happening, the decision of a task to do, and the planning of the behavior sequence, and thus will need much time and will accompany many erroneous behaviors. Hence, it is necessary to let passengers see the video in advance to prepare for the situation, so that the level of behavior should be changed from knowledge-based to rule-based. For rule-based behavior, the rule or what should be done in a specific situation is already stored in memory, thus passengers can remember the rule from memory and adequately follow the rule for wearing the life jacket. The lowered performance level during the overexcited situation can thus be supported by the demonstration video.

1.3.3.3 Physical Environment

The physical environment involves many aspects including lighting, sound, temperature, and space size. Regarding lighting, illuminance is recommended to be 750 to 1500 lx in the office. Usually, ICT devices such as a desktop computer, printer, and facsimile are used in such lighting conditions. But for mobile devices such as a laptop computer, smartphone, and wristwatch, the lighting condition will vary widely from 40,000 to 100,000 lx in the daytime to 100 to 300 lx in the evening. The visual display should adapt to the large dynamic range of illuminance. The environment where the user will use such devices cannot be limited to the optimum condition.

The sound environment will not affect the design of products as much compared to the lighting environment, except for a music-listening device and a device for the visually impaired. It is because most devices use visual information display and visual feedback, and do not rely much on the acoustic feedback.

The temperature of an office environment is usually regulated to 20°C (68.0°F) to 24°C (75.2°F) with a humidity of 45% to 65% in summer and 17°C (62.6°F) to 21°C (69.8°F) with humidity of 40% to 60% in winter. But we should not forget that there are users who have to work in a cold environment or a hot environment compared to an office. The optimum temperature for the computer and the human being is different. Hence, in the computer control room, the optimum temperature for the computer is a bit too cool for the human being (especially for women).

From the ergonomic viewpoint, the size of the work space in the office is recommended to be 8.5 to 8.8 m². Of course, it will vary depending on the type of task and the physical allowance. For example, the car driver is fixed by the seat belt in a small space. A more difficult situation is for passengers in the economy seat of an international flight who will have to endure the narrow space for 8 to 10 hours. Today, economy class syndrome is well known, but the physical environment has not yet improved for those who can afford only the economy seat.

1.3.3.4 Geographic Environment

When we think of the geographic environment, we tend to think of our own environment, but geographical environments vary so much from place to place on earth. There are physical differences, including weather, geological features, and water quality; biological differences, including plants, animals, insects, and pathogens; and social differences, including big towns, local towns, suburban areas, and deserted areas.

Manufacturers have been making efforts to let their products adapt to different geographic environments. One electronic company developed a vacuum cleaner for those living in the desert, who are annoyed by a big amount of small grains of sand. A new vacuum cleaner was developed with a large dust tank of 20 liters. Another chemical company developed a detergent for hard water, because people living in some areas of China found that the usual detergent did not clean clothes well with water containing calcium.

1.4 USER-CENTERED DESIGN

There are two similar terms: user-centered design (UCD) and human-centered design (HCD). The former was proposed by D.A. Norman in his 1988 book *The Psychology of Everyday Things* and the latter was proposed by ISO 13407 that was standardized in 1999 (now ISO9241-210).

1.4.1 User-Centered Design and Human-Centered Design

Human-centered design is a bit misleading because the word *human* can include all kinds of stakeholders, such as marketing people and manufacturers. If marketing people and manufacturers are included in "human," the design approach will be the one that focuses on the product and the service that will bring a big profit to them regardless of whether such artifacts are beneficial to the user.

Hence, the author distinguishes HCD and the UCD in the following sense:

HCD—The design approach that focuses on the humanity. An example is the systematic service approach for soldiers returning from war zones who suffer from posttraumatic stress disorder (PTSD). Another example is the design of social systems and/or the design of new restrooms for those who suffer from gender identity disorders. A futuristic example is the systematic approach for maintaining the dignity of human beings in the world of singularity. All these examples concern how the level of humanity can be improved. This should be the core of the HCD.

UCD—The design approach that focuses on users. Everybody in this society, even marketing people and manufacturers, can be users when they purchase something and use it. That is the moment when UCD should be considered. This is the core of the design approach of user engineering.

1.4.2 UI/UX

Figure 1.8 distinguishes the quality in design and the quality in use. (Figure 1.8 is the same as Figure 2.7 and more detail will be given in Chapter 2.) The bold outlined area on the left is the area for quality in design and on the right is the area for quality in use. The current combined buzzword UI/UX can be plotted on this figure, where UI is related to the design activity targeting the quality in design on the left, and UX is related to the quality in use on the right that is the quality of experience of using the designed artifact.

Although the expression UI/UX does not have a formal definition, the UI part can be interpreted as designing the user interface and the UX part can be interpreted as the resulting experience of users including expectation, purchase, and long-term usage. It should be noted that UXD (UX design) is an incorrect term. We can only say "designing for UX." UX is individualistic and subjective, and can be influenced by the user characteristics and the context of use, hence it is quite difficult to predict the resulting evaluation in advance. That is the reason why we cannot design the UX. We can only design for a better UX. In Figure 1.8, the designing phase is on the left side, and the right side represents the UX as a result.

1.5 CORE CONCEPT OF USER ENGINEERING

User engineering is engineering that aims to improve the quality of life of users by applying various concepts and methods of social science and human factors engineering. It does not target the sales promotion nor the development of simple novelty.

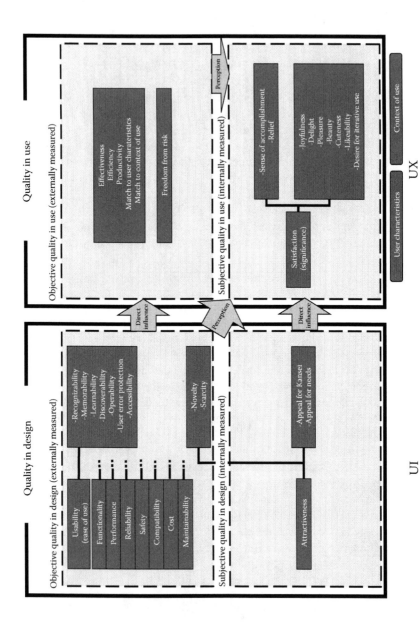

FIGURE 1.8 Quality in design and quality in use in relation to UI/UX.

It is true that consumers are sometimes attracted by the popularity and novelty. Regarding the popularity, people may feel a bit anxious if they might be regarded as old-fashioned and be left behind unless they purchase new artifacts. Novel products frequently enchant consumers for the simple reason that they are new. Indeed, it is a part of human nature to be enchanted by novelty. Hence, consumers will have to learn to be much wiser and to be more cautious so that they may not be misled by such factors.

But once they have purchased some products and have become users, they should be protected. One form of the protection is to establish the sales refund system for the simple reason "this product is not for me." There are some countries where this right is protected by the law, but in many countries, such as in Japan, it is not yet guaranteed. Users who have purchased wrong artifacts can only say the curse, the curse on themselves who purchased useless junk and the curse on the manufacturer who seized the money from them.

It is important that products and services should appear on the market not for sales but for the benefit of the user. In this world of commercialism, sales and profits have become the goal of corporate activity. But, from the viewpoint of user engineering, benefits on the side of users should be considered in the first place. Sometimes, it is true that users should not purchase products or services that are not fundamentally necessary for them and it might be much better for them to continue the conventional way of life, as is written in Chapter 12, Section 12.2.

2 Usability and User Experiences in the Context of Quality Characteristics

Usability has been the focus of attention of human factors specialists, cognitive psychologists, interface designers, and, of course, usability engineers for more than 30 years. At first, the concept was not clearly defined and was regarded as the same as the ease of use or usefulness. Until the 1980s, the approach from the cognitive psychology was rare and most studies on usability were conducted in the context of human factors engineering, and were focused on adaptability to physical characteristics, ease of operation, speed of operation, minimization of human error, fatigue, ease of perception, and so on.

But since the 1980s, when the personal computer appeared in the retail market, most of the use difficulty of such information technology (IT) devices and systems was regarded to have been caused by the lack of consideration of the cognitive characteristics of human beings. This situation has become more serious because of the lack of IT skills and the lack of necessary knowledge among such ordinary people as office managers, office workers, shopkeepers, students, and housewives. As Norman (1988) pointed out, ordinary users did not have the adequate mental model of the PC and interaction with the PC.

Discussion on the usability concept started in 1991 and the methodological preparation for measuring and evaluating usability also started at that time, and, as a result, usability engineering has established its own field in academia and industry. In parallel to the development of usability engineering, the concept of user experience (UX) appeared based on the idea that the usability concept was too narrow and thereafter the term UX was gradually used by more stakeholders. But some used the term *UX* just as the new aspect of the usability concept, and others emphasized the subjective aspects of UX too much. There was a history of confusion regarding the concept of UX, but it is true that the active discussion of UX involved many people from design, marketing, and advertising, as well as usability engineering, compared to the usability concept.

This chapter first describes the historical overview of the concept of usability and UX, and then proposes a general framework on the concept structure of quality characteristics that can explain both usability and UX.

2.1 CONCEPT OF USABILITY

Before UX became the buzzword, the focus of activity of human factors specialists and, later, usability professionals was placed on the concept of usability. In the following sections, the ideas of Shackel and Richardson, Nielsen, ISO 9241-11, Kurosu, ISO/IEC 9126-1, ISO 9241-210, and ISO/IEC 25010 will be introduced.

2.1.1 SHACKEL AND RICHARDSON (1991)

In academia, the concept of usability was first systematically defined by Shackel and Richardson (1991). As in Figure 2.1, they listed three important, positive aspects of artifacts: utility, usability, and likeability. Utility means that the artifact will do what is needed functionally. In other words, it is the functionality that matches the user's need. Usability means the degree of success that the user can work with the artifact. The success can be regarded as the same as goal achievement. Likeability, as coined by Shackel and Richardson, is similar to the subjective feeling of suitability, thus will lead to satisfaction. On the other hand, there are the negative costs of initial cost and running cost. The balance between the sum of utility, usability, likeability, and cost will affect the degree of acceptance, or the acceptability. If the former is equal to or larger to the latter, the artifact will be accepted and be purchased and used.

The significance of their model lies in usability being regarded as one of the important aspects of the artifact. But Shackel and Richardson did not specify the relative importance among utility, usability, and likeability. In other words, if an artifact has a high degree of utility and likeability, and the sum of the two will exceed the cost, there is a question if the artifact will be accepted even though it has a low level of usability. There must be a kind of absolute threshold for utility, usability, likeability, and cost. Furthermore, no other characteristics, such as performance,

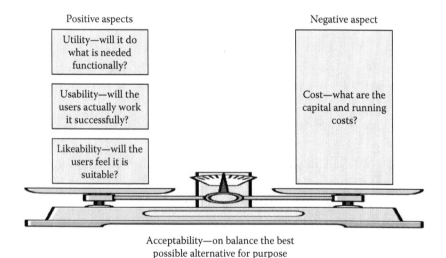

FIGURE 2.1   Conceptual model in terms of usability by Shackel and Richardson.

safety, reliability, and compatibility, are included. There is a question whether utility, usability, likeability, and cost are more important than performance, safety, reliability, and compatibility.

On the concept of usability, Shackel and Richardson asked, "Will the users actually work it successfully?" This question suggests the definition of the concept in ISO 9241-11:1998, which defined usability in the context of goal achievement where the success of goal achievement is measured by effectiveness and efficiency that finally lead to satisfaction.

2.1.2 NIELSEN (1993)

Two years later, Nielsen (1993) proposed a hierarchical model of usability in his book *Usability Engineering*. His model had two aspects: (1) the location of usability in the total set of quality characteristics and (2) subcharacteristics of usability.

Regarding the conceptual location of usability, Nielsen first made a distinction between usability and utility. This distinction is almost the same as the one by Shackel and Richardson, who regard utility to be the same as functionality. With regard to the concept of usability and utility, industry people had less concern about usability compared to utility. Even though users may be troubled by the lack of usability, they tend to be attracted by the functionality, hence engineers and managers in industry focused more on the development of new functions.

Utility and the usability are usefulness on the same level with such other quality characteristics as cost, compatibility, maintenance, reliability, and safety. They will lead to the practical acceptability of the product and further to the system acceptability. It is similar to Shackel and Richardson that the acceptability was placed as the ultimate stage of quality characteristics.

In addition, Nielsen specified the subcharacteristics of usability as learnability, efficiency, memorability, errors, and satisfaction. This is a unique aspect of Nielsen's concept structure and can be used as the criteria for the usability inspection using the heuristic evaluation method, which he also proposed in the book. Learnability, memorability, and errors are related to effectiveness, that is, whether the goal achievement will be successful, and such effectiveness-related concepts in addition to efficiency and satisfaction are the same as the idea of ISO 9241-11:1998 that was standardized 5 years later.

In Nielsen's model, the structure of quality characteristics is more systematically described and the location of usability is clearly specified. But we'll have to take care that Nielsen is also the one who proposed the heuristic evaluation method (see Chapter 9, Section 9.3.1). In other words, learnability and other characteristics located under usability are focal points when that method will be applied for evaluating the usability of artifacts. Learnability, for example, means to have fewer problems regarding learning. Likewise, with the exception of satisfaction, all characteristics under usability are proposed for detecting usability problems. That is to say, these characteristics are directing toward the zero level from the minus zone. On the contrary, the utility that can be presumed as consisting of functionality and performance is directing toward the plus zone from the zero level, because having some functionality or higher performance can be positively accepted by users.

This reflects the situation of usability engineering in the 1980s and 1990s when managers and engineers were more inclined to utility than usability. Furthermore, there are also the problems that the components of usability are not systematically chosen and do not cover all relevant characteristics. For example, the ease of cognition, including the visual size of the target and the contrast of the target against the background, is not included.

2.1.3   ISO 9241-11:1998

ISO 9241-11:1998 was standardized 5 years after Nielsen. In this standard, there is a formal definition of usability as the "extent to which a product can be used by specified users to achieve specified goals with effectiveness, efficiency and satisfaction in a specified context of use," where effectiveness is defined as "accuracy and completeness with which users achieve specified goals," efficiency is "resources expended in relation to the accuracy and completeness with which users achieve goals," and satisfaction as "freedom from discomfort, and positive attitudes towards the use of the product."

ISO 9241-11:1998 has been widely accepted as the standard definition of the usability concept especially in other ISO standards and documents such as ISO 13407:1999 (later revised as ISO 9241-210:2010); ISO/TR 16982:2002; ISO/TR 18529:2000; and ISO 20282-1:2006, ISO 20282-2:2006, ISO 20282-3:2007, and ISO 20282-4:2007. Although this standard does not refer to the relationship of usability to other quality characteristics, it is quite important that it specified the subcharacteristics of usability as effectiveness, efficiency, and satisfaction, with measures of such subcharacteristics by listing how they can be measured, and also specified the dynamic process of usability using the concept of the context of use including user, task, equipment, and environment.

In Annex B of ISO 9241-11, there is a list of measures for overall usability, desired properties of the product, and so on, with the measure for effectiveness that is mostly the number or the percentage for efficiency of time, and a rating scale for satisfaction. This list shows that effectiveness and efficiency as usability components can be objectively measured, whereas satisfaction can be subjectively measured.

In personal communication with the author in 2001, N. Bevan said the origin of the concept of usability in this standard was defined at the ISO TC159/SC4/WG5 meeting as "the degree to which specified users can achieve specified goals in a particular environment effectively, efficiently, comfortably and in an acceptable manner" and the word *satisfaction* was introduced for simplifying the definition as "freedom from discomfort and positive attitudes towards the use of the product" to have essentially the same meaning as the phrase "comfortably and in an acceptable manner."

The model of usability describes the dynamic process of how usability and its measures can be located. But we will have to be careful that usability measures are not mutually exclusive and collectively exhaustive (MECE) with one another. Especially, they are not independent of one another. First, efficiency cannot be measured when the goal is not achieved, hence efficiency is dependent on effectiveness. Second, satisfaction will be experienced when effectiveness and efficiency are satisfactory, and furthermore, it will be influenced by many other characteristics such as reliability, safety, and beauty. Hence, satisfaction should be regarded to be dependent of all these characteristics. Thus, the author adopted only effectiveness and efficiency as the measures of usability.

2.1.4   KUROSU (2005)

Within the scope of ISO 9241-11, I (Kurosu 2005a) proposed the model of goal achievement, as shown in Figure 2.2. This model describes effectiveness and efficiency in the context of goal achievement. The dotted line (ineffective) ends on the way to the goal and this suggests the occurrence of an error or the user being puzzled. It means that the use of an artifact is ineffective in this case. The dashed line, representing effectiveness, reaches the goal, but it takes a winding path thus is inefficient. It suggests that there was a trial and error. The straight line that reaches the goal in the shortest path means that it is effective and efficient. This figural representation only describes the usability concept proposed in ISO 9241-11. No other quality characteristics such as reliability, safety, and compatibility are described here.

2.1.5   ISO/IEC 9126-1:2001

In 2001, there appeared another ISO standard ISO/IEC 9126-1:1991, 2001 from JTC1. JTC1 is an ISO committee focusing on the software, whereas the ISO standards mentioned earlier were from TC159, a group of ergonomics specialists. Although ISO/IEC 9126-1:2001 was about the quality characteristics from the viewpoint of software development, it included a very important distinction between usability and quality in use. In this standard, the quality characteristics are divided into two groups, namely, software product quality (external and internal quality) and quality in use. The former

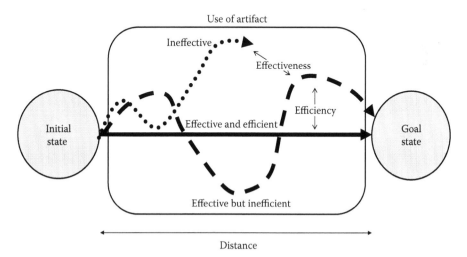

FIGURE 2.2   Model of goal achievement. (From Kurosu, M., 2005b, "Human Centered Design—Understanding of User and Evaluation of Usability," [In Japanese], HQL Seminar.)

includes functionality, reliability, usability, efficiency, maintainability, and portability, and the latter includes effectiveness, productivity, safety, and satisfaction.

By definition, external quality is "the totality of characteristics of the software product from an external view. It is the quality when the software is executed, which is typically measured and evaluated while testing in a simulated environment with simulated data using external metrics," and the internal quality is "the totality of characteristics of the software product from an internal view. Internal quality is measured and evaluated against the internal quality requirements." And quality in use is "the user's view of the quality of the software product when it is used in a specific environment and a specific context of use. It measures the extent to which users can achieve their goals in a particular environment, rather than measuring the properties of the software itself."

What should be noted is that usability is categorized as a part of the product quality (external and internal quality) and is different from the quality in use. The usability here includes understandability, learnability, operability, attractiveness, and compliance (to standards, conventions, style guides, or regulations relating to usability) as subcharacteristics. By comparing the notion of usability and quality in use, the former is the quality of the "ability" or the "potential" of the product, and the latter is the "result" of the use.

2.1.6   ISO 9241-210:2010

ISO 13407:1999 that adopted the usability definition of ISO 9241-11:1998 was standardized in 1999 and focused on human-centered design. In 2010, it was revised into ISO 9241-210:2010. The definition of usability in ISO 9241-210:2010 is expanded to include the "system, product or service" from that of ISO 9241-11:1998 that was applied only to the product. In other words, ISO 9241-210:2010 covers almost all kinds of artifacts, and the concept of UX is positioned as the goal of design instead of usability.

2.1.7   ISO/IEC 25010:2011

ISO/IEC 9126-1:2001 was later revised as a part of SQuaRE (Software product Quality Requirements and Evaluation) and was numbered as ISO/IEC 25010:2011. In this standard, the basic idea of ISO/IEC 9126-1:2001 is inherited to distinguish product quality and quality in use. The former includes functional suitability, performance, efficiency, compatibility, usability, reliability, security, maintainability, and portability, whereas the latter includes effectiveness, efficiency, satisfaction, freedom from risk, and context coverage.

What is distinctive in this revised version in terms of usability is that usability includes appropriateness, recognizability, learnability, operability, user error protection, user interface aesthetics, and accessibility. This list of subcharacteristics is a bit different from that of Nielsen and ISO/IEC 9126-1:2001 but includes all possible abilities and potentials that the product will have to fulfill during the design phase. Another feature of this revised version is that the subcharacteristics of usability in ISO 9241-11:1998, that is, effectiveness, efficiency, and satisfaction, are all excluded from usability and are included in quality in use. This is quite confusing but is clearer in the sense that these three subcharacteristics are specific to the real context of use

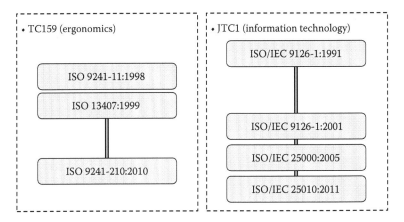

FIGURE 2.3 ISO standards related to usability.

and are subject to change from usage to usage. In other words, the subcharacteristics of usability in ISO 9241-11:1998 are all related to the quality in use, although they were previously positioned under the name of usability.

The concepts of usability and quality in use of ISO/IEC 25010:2011 are claimed to be applied only to the software, but I think that this idea can be applied to all artifacts including hardware, software, and humanware (service) with minor modifications.

2.1.8 A Note on ISO Standards in Terms of Usability

Because the system of ISO standards is a bit complex, additional information will be given here as in Figure 2.3. There are two TCs (Technical Committees) that relate to the concept of usability: TC159, which concerns ergonomics, and JTC1, which concerns information technology. From TC159, there appeared ISO 9241-11:1998, ISO 13407:1999, ISO 18529:2000, ISO/TR 16982:2002, ISO/PAS 18152:2003, ISO 20282-1:2006, ISO/PAS 20282-2:2013 (revises ISO/PAS 20282-3:2007 and ISO/ PAS 20282-4:2007), and ISO 9241-210:2010. From JTC1, ISO/IEC 9126-1:1991, ISO/IEC 9126-1:2001, ISO/IEC 25000:2005, and ISO/IEC 25010:2011 were standardized. From the author's view, standards from TC159 are full of suggestions and those from JTC1 are more systematic.

2.2 CONCEPT OF USER EXPERIENCE (UX)

In 1993, Norman joined Apple Computer with the job title of "user experience architect." In 1998, Norman formally proposed the concept of UX (user experience). He said, "I invented the term because I thought human interface and usability were too narrow. I wanted to cover all aspects of the person's experience with the system including industrial design, graphics, the interface, the physical interaction, and the manual." But, in 2007, he said, "Since then the term has spread widely, so much so that it is starting to lose its meaning" and "User experience, human centered design, usability, even affordances just sort of entered the vocabulary and no longer have any special meaning. People use them often without having any idea why, what the word means, its origin, history, or what it's about" (Norman and Merholz 2007).

This statement of Norman summarizes the birth and the current chaotic situation of UX. According to All About UX (www.allaboutux.org/), there are (at least) 27 different definitions of the concept. They include

- All aspects of the end-user's interaction with the company, its services, and its products. (Nielsen-Norman Group)
- Every aspect of the user's interaction with a product, service, or company that make up the user's perceptions of the whole. (UPA, later UXPA)
- A person's perceptions and responses that result from the use or anticipated use of a product, system or service. (ISO 9241-210:2011)
- A consequence of a user's internal state (predispositions, expectations, needs, motivation, mood, etc.), the characteristics of the designed system (e.g. complexity, purpose, usability, functionality, etc.) and the context (or the environment) within which the interaction occurs (e.g. organizational/ social setting, meaningfulness of the activity, voluntariness of use, etc.) (Hassenzahl and Tractinsky 2006)

In order to cope with this chaotic situation, a gathering of thirty UX specialists was held in Dagstuhl in 2010 and the result of the discussion was summarized as the "User Experience White Paper" in 2011 (Roto et al. 2011).

2.2.1 Temporal Aspect of UX

The long-term temporal sequence of UX is one of the major focuses of the white paper. The white paper distinguishes four types of experience, namely, anticipated UX, momentary UX, episodic UX, and cumulative UX.

Including anticipated UX in the whole set of UX is very important. Users integrate their past experiences for similar products/services as well as the information obtained from various sources such as a catalog, corporate website, magazine, newspaper, and personal communication. The integration of this information will form the expectation among users who have not yet used the products/services, hence they should better be called as consumers than users. Anyways, the expectation is surely a part of the total experience.

Momentary UX and episodic UX differ only in their time span and seem to be regarded as the same in many situations. These are elementary experiences during long-term use until the end of use that may result from many reasons including performance deterioration, damage, and loss of interest.

Cumulative UX is not necessarily the experience at the final stage of use, but the experience that can be addressed by the user at any time as the current impression. The experience value at that time will reflect the average value of past experiences with exponential weight, that is, the more recent is more influential. Thus, cumulative UX is the UX value at the time of measurement.

Kujala et al. (2011) showed that the evaluated value in the UX curve goes up and down dynamically during long-term use. Thus, the value of evaluation of UX may change even after the evaluation at a certain point. In other words, UX cannot be represented as a single summative value but should be regarded as the curve that is dynamically changing. Hence, seeking a single value that will represent the UX may not be adequate. Of course, when asked, users will give the cumulative UX as a single value, for example, on a 7-point rating scale. But we'll have to be careful that the value is only valid at the time when the question was asked. That value may go up or down afterward based on future use.

A psychological theory of adaptation theory proposed by Helson (1964) is closely related to the concept of UX. His idea can be represented as follows:

$$\log AL = \sum_{i=1}^{n} wi \cdot logAi \quad \left(\sum_{i=1}^{n} wi = 1 \right)$$

where n is the number of past events that influence the current judgment, wi is the weight for the ith past event, and Ai is the strength of the ith past event. This equation means that our judgment (on the experience) will be influenced by the adaptation level (AL) of which the logarithmic of AL is a weighted logarithmic sum of past experiences. And the current judgment (J) for the current event (in this context, the experience) can be represented as follows:

$$J = k \, (\log X - \log AL)$$

This means that the judgment J (of an experience) is a weighted difference between the strength of the current value (of experience) and the adaptation level. If the value of X is larger than AL, then J will be positive and vice versa.

To take an example from everyday UX examples, a traveler who usually stays in deluxe hotels will have a high AL and, thus, will evaluate his experience of staying in a modest B&B as an unpleasant experience, whereas another traveler who has been hitchhiking will have a relatively low AL and will evaluate his experience of staying in the same B&B as a pleasant experience. In such a manner, our evaluation of the UX may vary depending on past experiences.

2.2.2 SUBJECTIVE ASPECT OF UX

Another characteristic of UX is that the UX is influenced not only by the objective quality characteristics such as usability, functionality, and safety, but also by the subjective quality characteristics such as pleasure, joyfulness, and, more important, satisfaction.

Although ISO 9241-11:1998 provided the definition of usability as composed of effectiveness, efficiency, and satisfaction and is widely accepted today, I have been opposed to this definition at ISO TC meetings since the 1990s, because satisfaction should not be locally positioned as a part of usability but should be located at a higher position for the reason that it is influenced not just by effectiveness and efficiency. It is influenced by many other quality characteristics including reliability, safety, functionality, and aesthetic aspects. Satisfaction is the utmost quality characteristic and belongs to the subjective aspect, thus is much related to UX (Kurosu 2005b).

There are many researchers who put emphasis on the importance of the subjective aspect. Jordan (1998, 2000) proposed that pleasure is quite important as well as functionality and usability. His concept is structured in a similar way as the psychologist Maslow who proposed a hierarchical structure of human needs consisting of physiological needs, safety needs, belongingness and love needs, esteem needs, and self-actualization needs from the bottom to the top. The hierarchical structure of user needs proposed by Jordan is from functionality, through usability, and reaches pleasure at the top. He further classified the pleasure into four types: physical, social, psychological, and ideological. It is based on the idea of the anthropologist Tiger. Physical pleasure is the pleasure derived from the sense organs including pleasures connected with touch and smell and is related to sexual pleasure. Social pleasure is the enjoyment derived from the company of others. Psycho-pleasure can be gained from accomplishing a task, thus is closely related to the usability concept. Finally, ideological pleasure is the pleasure derived from theoretical entities such as a book, music, and art, and is related to aesthetics.

Hassenzahl (2003) differentiated the pragmatic attributes and the hedonic attributes. The former consists of the utility (i.e., the functionality) and the usability (i.e., the way to access the functionality). The latter is the affective quality of a product including pleasure, enjoyment, and fun derived from possession or use of a product.

Norman (2004) emphasized the emotional aspects of a product. He wrote, "What many people don't realize is that there is a strong emotional component to how products are designed and put to use. I argue that the emotional side of design may be more critical to a product's success than its practical elements."

The aforementioned authors emphasized the importance of the subjective aspects of UX regarding products, systems, and services, in addition to such objective aspects as utility and usability.

2.3 A MODEL OF QUALITY CHARACTERISTICS

In this section, a new model of quality characteristics is proposed based on the critical review of past theories and models. This model includes two dimensions where one distinguishes the quality in design and the quality in use, and another distinguishes the objective quality characteristics and the subjective quality characteristics.

2.3.1 QUALITY IN DESIGN AND QUALITY IN USE

The basic idea of this distinction comes from ISO/IEC 25010:2011 in which the product quality and the quality in use are distinguished. But product quality is the result of the design process where engineers and designers strive for achieving the quality characteristics to be sufficiently enough. Hence, the name should better be the "quality in design" rather than the "product quality" considering the activities of engineers and designers. The quality in use is a good name because it specifies the quality while the product/service is used. As a result, the relationship between the quality in design and the quality in use is somewhat similar to the causal relationship where the quality in design is the cause and the quality in use is the result.

From this viewpoint, the conceptual framework by Shackel and Richardson (1991), Nielsen (1993), and ISO 9241-11:1998 can be said to be ambiguous. It is not clearly stated whether the concept of usability concerns the usability in design or the usability in use. This point can be clarified in the model proposed next (Kurosu 2015a).

2.3.2 Model of Quality Characteristics

Quality control has established the procedure for clarifying the criteria regarding the externally observable quality characteristics. Measures used for this purpose are duration, frequency of occurrence, or some numerical measures resulting from a specifically designed formula. Such objective measures are useful for making a decision in business situations.

But, in addition to these objective quality characteristics, subjective aspects such as likeability and satisfaction should not be forgotten, because they were included in the model on usability by Shackel and Richardson, Nielsen, and ISO 924-11:1998. Since UX has become the focal point of design and development, subjective aspects such as pleasure, hedonic attributes, fun, and joyfulness have become the center of concern of practitioners.

Furthermore, UX is an individual and internal experience, as was described in the UX white paper. Subjective quality characteristics have been mostly measured by psychological rating scales, but applying physiological methods such as the eye camera, electroencephalogram (EEG), galvanic skin response (GSR), electrocardiogram (ECG), near-infrared spectroscopy (NIRS), and functional Magnetic Resonance Imaging (fMRI) are now becoming popular in addition to the psychological measurements even though the conceptual validity of each physiological measure is difficult to confirm.

2.3.3 Four Regions of Quality Characteristics

By combining these two dimensions of quality characteristics, i.e., "quality in design" vs. "quality in use," and "objective quality" vs. "subjective quality," in Figures 2.4 and 2.5, there will be

- Objective quality in design
- Objective quality in use

FIGURE 2.4 Quality in design and quality in use.

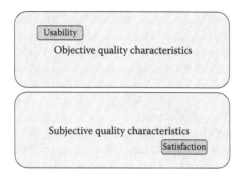

FIGURE 2.5   Objective quality characteristics and subjective quality characteristics.

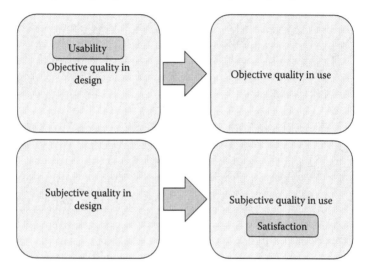

FIGURE 2.6   Four areas of quality characteristics.

- Subjective quality in design
- Subjective quality in use

as shown in Figure 2.6. Details on the quality characteristics in these four regions are shown in Figure 2.7.

2.3.4   OBJECTIVE QUALITY IN DESIGN

Objective quality in design is similar to the product quality in ISO/IEC 25010:2011 and includes the following qualities (definition of each quality characteristic is different from ISO/IEC 25010:2011):

Usability (ease of use)—The capability of an artifact to be used with ease. This concept is almost the same as the concept of ease of use. Subcharacteristics of this quality characteristic will be defined later.

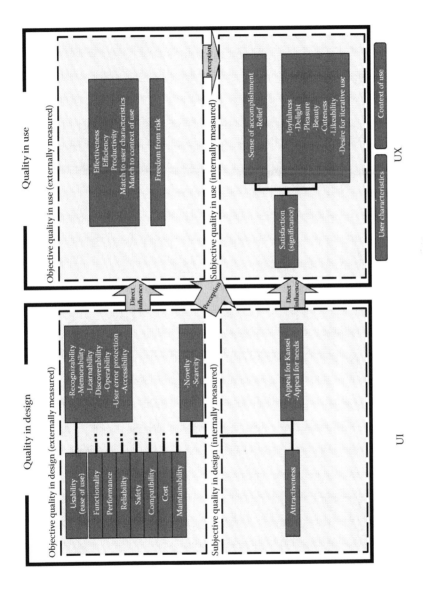

FIGURE 2.7 Four areas of quality characteristics: objective quality in design, objective quality in use, subjective quality in design, and subjective quality in use.

Functionality—The ability of an artifact to support the user functionally to achieve the goal. It is sometimes called the utility in combination with the performance.

Performance—The ability of an artifact to help users achieve the goal with shorter time, with less consumption of energy, and so on.

Reliability—The capability of an artifact to maintain the designated level of performance regardless of the disturbance.

Safety—The capability of an artifact not to harm the user or the property of the user.

Compatibility—The capability of an artifact that the functionality can be used in a different environment.

Cost—Monetary expense for an artifact including the initial cost and the running cost. The initial cost is the cost for obtaining the artifact, and the running cost is the cost after the purchase so that the artifact can work constantly thereafter.

Maintainability—The capability of an artifact to allow maintenance to be easily done.

In Figure 2.7, subcharacteristics of usability can be defined as follows:

Recognizability—Ease of recognition that can be achieved when the artifact is designed so that it can fit the nature of human cognitive mechanism such as pattern recognition, integration of information, and decision making.

Memorability—Ease of memorizing that includes the encoding to memory, the retention in memory, and the retrieval from memory in terms of the name of function, procedure, and expected result. It should be noted that the recognition (e.g., menu selection interface) is easier than the recall (e.g., command interface).

Learnability—Ease of learning the procedure for using an artifact. Usually, the performance of learning takes an S-shaped curve in terms of the time or the amount of exercise, where the performance goes up slowly in the early stage, then goes up until the curve reaches the saturation level.

Discoverability—Ease of discovering clues to use the functionality. Clues include hardware components such as switches, buttons, levers, and meters, and software components such as windows, menus, and icons, as well as procedures for achieving the goal.

Operability—Ease of operation. Usually, the shorter the steps of operation, the easier it is to perform the operation. Operability also concerns the total length of the operational path.

User error protection—Providing the information to confirm what was done to the system, and providing an undo and redo function.

Accessibility—There are two different interpretations to this concept. One is that accessibility is the ease of access to the artifact before using it ("access" the artifact, then "use" it). In this interpretation, accessibility is different from usability. Another interpretation is that accessibility is the usability for a disabled user.

Of course, all other quality characteristics, such as functionality and performance, have subcharacteristics, as in the case of usability of which the reader should refer to ISO/IEC 25010:2011.

One important point here is that all these quality characteristics are still the potential of an artifact. It can be seen from their names; many of them have the suffix -*ability*. Ability or capability is the potential of an artifact, and the user will be guided toward the goal by such ability. However, ability is different from the result. Ability does not guarantee a successful result. A simple example is that even a talented student may not score 100% on an examination, where the talent is the ability and the examination is the result.

In addition to the aforementioned quality characteristics, two different types of quality in design are included here: novelty and scarcity.

Novelty—The degree of recency in the market. By using terms in the theory of diffusion of innovations by Rogers (2003) (see Chapter 1), the market can be divided into five groups of customers, namely, innovators, early adopters, early majority, late majority, and laggards based on the tendency to be attracted by new innovative products and services. Furthermore, according to Moore (2014), there is a chasm between early adopters (opinion leaders) and the early majority, and the innovative products and services can be a de facto standard if they can come across the chasm, otherwise they will disappear from the market. The novelty can be objectively measured in time units (e.g., months) after the release and will affect the attractiveness that is located in the subjective quality in design.

Scarcity—The degree of relative minority in the market. Some maniac users are attracted to a product for the reason that it is not easily found in the market. Scarcity can be objectively expressed in the total number of products released in the market and will influence the attractiveness.

These two measures are objective quality in use because they can be counted or measured numerically. But they will influence the attractiveness in the subjective quality in design instead of the objective quality in use.

2.3.5 OBJECTIVE QUALITY IN USE

Basically, the objective quality in use is almost the same as the quality in use of ISO/IEC 25010:2011. Especially, the definitions of effectiveness and efficiency come from ISO 9241-11:1998. Although effectiveness and efficiency look similar to the subcharacteristics of usability in objective design quality, they are not the "ability" but are the result of use of the artifact by a specific user in a specific context of use.

Effectiveness—According to ISO 9241-11:1998, effectiveness is the "accuracy and completeness with which users achieve specified goals." This can be measured by those externally observed measures such as "the percentage of goals achieved, the percentage of users successfully completing task, and the average accuracy of completed task" (Table B.1 of ISO 9241-11:1998).

Efficiency—"Resources expended in relation to the accuracy and completeness with which users achieve goals" (definition of ISO 9241-11:1998). This
also can be measured by those externally observed measures such as "the
time to complete a task, the tasks completed per unit time, the monetary
cost of performing the task" (Table B.1 of ISO 9241-11:1998).

Productivity—The level of performance of an artifact to enable users to expend
appropriate amounts of resources in relation to the effectiveness achieved in
a specified context of use (adapted from ISO/IEC 9126-1:2001). Although
productivity is not included in ISO/IEC 25010:2011 (as the revision of ISO/
IEC 9126-1:2001) and is difficult to measure numerically, this is an important aspect of the objective quality in use.

Match to user characteristics—The concept of user is clearly defined in ISO/
IEC 25010:2011 where the direct user includes the primary user and the
secondary user, where the former is defined as "a person who interacts with
the system to achieve the primary goals" and the latter is defined as "those
who provide support," and the indirect user "receives output but does not
interact with the system." The definition of a user in ISO 9241-11:1998 is
just "a person who interacts with the product." In other words, ISO 9241-
11 only refers to the primary user based on the definition of ISO/IEC
25010:2011. In almost all design processes, the main targeted user is the
primary user. For example, personas are frequently used in today's product/
service design, but in most cases they describe only a limited number, just a
few or a several, of the primary users as personas. Regarding UX, the customer journey map is used frequently in terms of the primary user. Primary
users are described in both of the "to-be" customer journey map during the
design phase and the "as-is" customer journey map during the usage phase
of a product/service. And as has been discussed in Chapter 1, Section 1.3,
the user has three different aspects: characteristics, inclination and situation
and environment. Hence, there is a large diversity among users. This is the
main drive for the movement of universal design or design for all.

Match to context of use—The definition of context of use in ISO 9241-11:1998
is "users, tasks, equipment (hardware, software and materials), and the
physical and social environments in which a product is used." But the
author removes users from the context of use, because users are not the part
of the context but are situated in the context (of use).

Freedom from risk—"The degree to which a product or system mitigates the
potential risk to economic status, human life, health, or the environment"
(ISO/IEC 25010:2011).

It is a bit strange that the subcharacteristics of usability in ISO 9241-11:1998—
effectiveness, efficiency, and satisfaction—are included here as the subcharacteristics of quality in use. It is because of this fact that ISO 9241-11 does not distinguish
usability and quality in use, or more generally, product quality and quality in use. In
fact, the concepts of effectiveness and efficiency are not the part of usability, which
is a word combining *use* and *ability*. Effectiveness and efficiency depends on the user

characteristics and the context of use, and thus are the parts of quality in use for a user with a certain set of characteristics and in a specific context of use.

Another point that should be noted regarding this category is that satisfaction is not included here unlike in ISO 9241-11:1998 and ISO/IEC 25010:2011. It is because satisfaction is not a part of usability or objective quality characteristics but is a part of the subjective quality characteristics.

There is a link between objective quality in design and objective quality in use that shows the natural relationship from design to usage.

2.3.6   SUBJECTIVE QUALITY IN DESIGN

Subjective quality in design can be summarized as the attractiveness that includes the appeal for Kansei and the appeal for needs as well as the novelty and scarcity.

Kansei is a mental process related to emotion and perception. Kansei engineering (affective engineering), which originated in Japan some 50 years ago, has been dealing with such phenomenon as beauty and cuteness. Kansei experience is related to positive emotions including joy, pleasantness, and delight, which ultimately lead to satisfaction. These phenomena and emotions are part of subjective quality in use because they are Kansei experiences or experiences related to emotion, and are perceived by the user in the real context of use. Appeal for Kansei is a candidate or a potential for subjective quality in use, thus it is a part of subjective quality in design. Designers strive to implement positive subjective elements in their design work so that the user can have a positive Kansei experience. But their efforts do not always result in a positive experience. There is no guarantee that users may have the positive Kansei experience as expected.

Aesthetical impressions such as beauty and cuteness are interesting because people usually regard them as part of the attribute of the object, but actually the object itself is just the physical object without any such subjective impression. These impressions are the result of the psychological mechanism of projection, as can be seen in projective methods used in clinical psychology. Emotional states of users are projected onto the artifact. Thus, the psychological mechanism of projection is not the ordinary perception but should be called an apperception, the term used in clinical psychology especially in relation to the projective method. Hence, the elements for beauty and cuteness as the quality in design are just a possibility for the Kansei experience or the quality in use.

> Appeal for needs—In everyday life, people have some needs and means for fulfilling them. If designers can foresee what kind of needs the user may have, the design process will become much easier. Frequently the needs of the user are covert, and market research and business ethnography are conducted for the purpose of understanding user needs. Market research and business ethnography will give some information on user needs, but they cannot be fully valid because the user may not completely understand their own needs for themselves in everyday life. Because of this reason, designing for users' needs is a risky challenge, and designers make efforts to ensure the product/service to be designed has full potential to be accepted.

Similarly, in the case of Kansei, designing for users' needs is a challenge to conform the product/service to users' needs.

Appeal for Kansei—See Section 2.4.1.

Attractiveness—The appeal for Kansei and the appeal for needs as well as novelty and scarcity form the attractiveness as the total measure of subjective quality in design. The user will be attracted to the product/service in terms of such characteristics, but, in some cases, will not be attracted by such characteristics. This is the difficulty of designing.

2.3.7 Subjective Quality in Use

Subjective quality in use is a group of quality characteristics that represent the subjective impression during the use of a product/service. It includes a sense of accomplishment and relief that are related to the objective quality in use. In addition, there are joyfulness, delight, pleasure, beauty, cuteness, likeability, and desire for iterative use that are related to the subjective quality in design.

Sense of accomplishment—When users accomplish the task and achieve the goal, they may have a positive feeling and get relaxed. Usually, the greater the task difficulty, the greater the sense of accomplishment. This is related to the objective quality in use and when it is perceived that the goal was achieved, the user will feel this sense, at least, to some extent.

Relief—When the task is accomplished, users usually feel relief as well as a sense of accomplishment. This subjective quality in use is related to the perception of the objective quality in use.

Joyfulness—Joyfulness is an example of the positive subjective quality in use. The user may feel joyfulness when the task is successfully completed. The term *flow* by Csikszentmihalyi (1990) and *hedonic attributes* by Hassenzahl (2003) may be categorized as attributes containing this subjective quality in use. Of course, the goal achievement may be successful but sometimes may not be reached. Hence, negative feelings that are contrary to those listed here should be considered. In Figure 2.7, joyfulness, delight, and pleasure belong to the emotion, while beauty and cuteness belong to the aesthetical judgment or Kansei.

Delight—Delight is also a positive feeling and can be experienced when something of great joy is obtained. Likewise, other descriptive words for positive feelings, such as gratification, happiness, and gladness, can be added into this list.

Pleasure—Pleasure should be kept on the list of positive feelings because Jordan used this term frequently. It is true that something more should be considered in terms of the quality of the product/service in addition to objective quality characteristics such as functionality and usability and that should be a subjective one.

Beauty—The impression of beauty, as was explained earlier, is the subjective quality in use that should correspond to the appeal for Kansei. Historically, beauty is a very difficult issue as it leads to questions of what and why something is beautiful. Aesthetics has been dealing with this issue for a long time. It is because the concept of beauty changes from time to time and from place to place and even differs from individual to individual. The concept of beauty is quite individualistic for a person who lives in a specific time and place with the influence of a specific culture. Many people can be impressed by *Mona Lisa* but some others may be impressed by *Fountain* by Marcel Duchamp. Some people may feel that the lemon squeezer by Alessi is beautiful, but some others may dislike it because it looks like a bug or an alien. An artifact can be judged to be beautiful, but at the same time it cannot be judged to be beautiful depending on the internal Kansei structure. There is no guarantee that an artifact will be judged beautiful in the phase of quality in use even though the design is done deliberately in the phase of designing.

Cuteness—This Kansei impression is somewhat similar to beauty. Take the example of a stuffed doll. Many stuffed dolls have soft shaggy skin and are round shaped with big eyes and smiling curved mouths as the elements of subjective quality in design or an appeal for Kansei. But the response of the customer may be dispersed from positive to negative. It is difficult to predict if the product will be accepted as cute in the market until it is released and used.

Likeability—This is the term used by Shackel and Richardson for proposing something different from the objective quality characteristics of functionality and usability.

Desire for iterative use—In the system usability scale (SUS) (Brooke 1996), there is an item "I think that I would like to use this system frequently," and in the definition of satisfaction in ISO 9241-11:1998 there is the phrase "positive attitudes towards the use of the product." It is clear that the word *use* means "reuse" in these cases. As such, the desire for the reuse or the iterative use of the product/service should be regarded as a part of the subjective quality in use and be included here.

Similar to the relationship between the objective quality in design and the objective quality in use, there is a link between the subjective quality in design and the subjective quality in use as a natural relationship from the design to the usage.

There are two other links: one is from the objective quality in design to the subjective quality in use and another is from the objective quality in use to subjective quality in use. There is not a direct link between them, but the perception of objective quality in design and objective quality in use influence the subjective quality in use. This point will be discussed in the next section.

2.4   SATISFACTION

According to the *Oxford English Dictionary, satisfaction* is defined as "the action of gratifying (an appetite or desire) to the full, or of contenting (a person) by the fulfilment of a desire or the supply of a want. The fact of having been thus gratified or contented." The important keyword here is *full*. In other words, there is a certain volume of space to be filled in the human need or want. People recognize something as attractive when it seems to fill their need or want. And they tend to try to do something to get it so that the space will be filled. That is the motivation mechanism of human beings. The space could be filled in terms of many quality characteristics including both objective (usability, reliability, etc.) and subjective (novelty, scarcity, beauty, cuteness, etc.) ones.

The mechanism of need fulfillment is multiplication rather than addition. Take the example of objective quality characteristics and subjective quality characteristics. In the additive model, the lack on one aspect can be conpensated by another aspect so that the sum of the two will exceed the threshold for acceptance. But this is not reality. The lack of subjective quality characteristics, for example, cannot be filled by the high level of objective quality characteristics. Instead, the mechanism is more multiplication. The low level (e.g., $q = 0.3$, where $0 \leq q \leq 1$) on one side cannot be supplemented by the high level (e.g., $q = 0.8$) on the other side, thus $0.3 \times 0.8 = 0.24$. Even if it is not multiplication, the minimum rule can be applied to give the result of $q = \min(0.3, 0.8)$. This kind of logic can be viewed in Kano's theory of attractive quality (1984).

2.4.1 KANSEI

The term *Kansei* comes from the Japanese word 感性, which has a bit of a vague definition. It is related to sensation, perception, cognition, and emotion at the same time. For example, the simple word *bright* will have at least two different meanings: one is the physical state of the light source reflecting or emitting the light, and another is the somewhat metaphorical meaning of the mental state of liveliness and cheerfulness. While the former is related only to the sensation, the latter has some psychological meaning and is related to Kansei.

Another example is when we look at the dark clouds in the evening when we are driving in a deserted area. Usually, the driver will think that there are dark clouds and rainfall may come. This is the case of the usual perception. But drivers who have anxiety or are superstitious may feel that something bad might happen (Figure 2.8).

FIGURE 2.8   Objective observation and subjective observation.

This is an example of projection, as discussed later, and is an example of a Kansei experience.

Readers with a background in clinical psychology may recognize that the concept of apperception can be found in projection methods such as the Rorschach test or the thematic apperception test (TAT). In a metaphorical sense, we project our mental state onto the external stimulus and interpret it based on our thoughts and feelings. When we think of a stuffed bear as cute, the cuteness is not the physical characteristics of the doll but our projected feeling onto the doll. This is apperception.

In Kansei engineering, such apperceptual characteristics are regarded as expressing Kansei in addition to such perceptual characteristics as noisy, bright, and heavy. In short, Kansei is regarded as the result of perception in some cases, but, in most cases, it is the result of apperception based on the coordination of cognition and emotion. It is the basic psychological mechanism for subjective quality characteristics.

Behind such phenomena, a psychological process model as shown in Figure 2.9 can be assumed. Gray areas show the information processing model of human cognition that is generally accepted today. In addition, there is the emotional processing model shown in black and dark gray with bold dashed lines. It has not yet been physiologically confirmed, but as a functional level, we can think of information and valence (emotional value) as a unit for processing. The information taken from the external world evokes a certain level of valence and, at the same time, retrieves

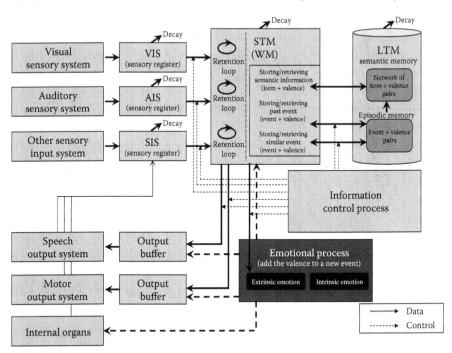

FIGURE 2.9 Psychological process model of information and emotion. (From M. Kurosu, 2010, "A Tentative Model for Kansei Processing—A Projection Model of Kansei Quality," Proceedings of Kansei Engineering & Emotion Research [KEER] 2014 International Conference.)

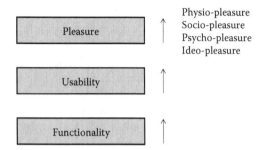

FIGURE 2.10   Jordan's idea.

pairs of information from long-term memory (LTM). As a result of some calculation in terms of the value of valence at short-term memory (STM) or working memory (WM), the resulting valence for the incoming information will be determined and then stored in the LTM.

2.4.2   Jordan (1998)

Jordan (1998, 2000) proposed a three-layered hierarchical model composed of functionality, usability, and pleasure, as is described in Figure 2.10 (also refer to Section 2.2.2). According to Jordan, functionality is the fundamental characteristics for a product. But usability is also quite important so that the function will be used (effectively and efficiently). His idea is more than that. He put pleasure above functionality and usability, because it makes the product attractive. Then he differentiated four types of pleasure, namely, physio-pleasure, socio-pleasure, psycho-pleasure, and ideo-pleasure.

Jordan's concept of pleasure as a subjective quality characteristic is similar to the notion of satisfaction in the ideas of Nielsen and ISO 9241-11, but is different from them in that it is differentiated from usability and is an independent concept. In this sense, his idea is more similar to that of likeability by Shackel and Richardson, but is more marked as being positioned above other quality characteristics.

2.4.3   Hassenzahl (2004)

A clear differentiation between objective quality characteristics and subjective quality characteristics was made by Hassenzahl (2004). Using his terminology, he distinguished pragmatic attributes from hedonic attributes. Hedonic attributes is his unique terminology but has something in common with subjective quality characteristics.

2.4.4   Satisfaction as the Ultimate Concept

In order to confirm the dependency among quality characteristics, Kurosu and Hashizume (2014) adopted conceptual dependency analysis. With this method, two arbitrary concepts are first selected. Then the following questions are asked: Can

Category	Distance	Dependency	Relationship
1	0	A then B B then A	Identical
2a	1	A then B or not B B then A	B is dependent on A
2b	1	A then B B then A or not A	A is dependent on B
3	2	A then B or not B B then A or not A	Partial dependence
4	3	A then not B B then not A	Independent

FIGURE 2.11 Dependency patterns.

A \ B	Functionality	Performance	Ease of cognition	Ease of operation	Effectiveness	Efficiency	Reliability	Cost	Safety	Compatibility	Maintenance	Pleasure	Joy	Beauty	Attachment	Motivation	Value	Meaningfulness	Satisfaction
Functionality	0	3	2	2	2	2	3	3	3	3	3	3	3	3	3	3	3	3	1
Performance	3	0	2	2	2	2	3	3	3	3	3	3	3	3	3	3	3	3	1
Ease of cognition	2	2	0	3	2	2	3	3	3	3	3	3	3	3	3	3	3	3	1
Ease of operation	2	2	3	0	2	2	3	3	3	3	3	3	3	3	3	3	3	3	1
Effectiveness	2	2	2	2	0	1	3	3	3	3	3	3	3	3	3	3	3	3	1
Efficiency	2	2	2	2	1	0	3	3	3	3	3	3	3	3	3	3	3	3	1
Reliability	3	3	3	3	3	3	0	2	2	3	3	3	3	3	3	3	3	3	1
Cost	3	3	3	3	3	3	2	0	3	3	3	3	3	3	3	3	3	3	1
Safety	3	3	3	3	3	3	2	3	0	3	3	3	3	3	3	3	3	3	1
Compatibility	3	3	3	3	3	3	3	3	3	0	2	3	3	3	3	3	3	3	1
Maintenance	3	3	3	3	3	3	3	3	3	2	0	3	3	3	3	3	3	3	1
Pleasure	3	3	3	3	3	3	3	3	3	3	3	0	0	2	3	2	2	2	1
Joy	3	3	3	3	3	3	3	3	3	3	3	0	0	3	3	2	2	3	1
Beauty	3	3	3	3	3	3	3	3	3	3	3	2	3	0	2	2	2	3	1
Attachment	3	3	3	3	3	3	3	3	3	3	3	3	3	2	0	3	3	3	1
Motivation	3	3	3	3	3	3	3	3	3	3	3	2	2	2	3	0	2	2	1
Value	3	3	3	3	3	3	3	3	3	3	3	2	2	2	3	2	0	2	1
Meaningfulness	3	3	3	3	3	3	3	3	3	3	3	3	3	3	3	2	2	0	1
Satisfaction	1	1	1	1	1	1	1	1	1	1	1	1	1	1	1	1	1	1	0

FIGURE 2.12 Result of concept dependency analysis (distance matrix).

concept A always be approved if concept B is achieved? Can concept B always be approved if concept A is achieved? Based on the results, five types of relationships can be determined, as shown in Figure 2.11. For example, if the artifact is reliable, it can be said to be satisfactory. But even if it is satisfactory, it is not always true that the artifact is full of reliability. This analysis resulted in the fact that satisfaction is the topmost concept among all those quality characteristics listed earlier, as shown in Figure 2.12.

2.4.5   Kano's Theory of Attractive Quality (1984)

Kano et al. (1984) distinguished "attractive quality" and "must-be quality" in relation to satisfaction and needs fulfillment. Satisfaction and needs fulfillment are mutually independent.

Must-be quality will dissatisfy users when it is not fulfilled. Even when it is fulfilled, it does not give users a high level of satisfaction, just non-dissatisfaction. Hence, the maximum level of must-be quality is at the neutral level (zero level) in terms of satisfaction (dissatisfied versus excitement) and the must-be quality will not reach the zone of excitement.

On the contrary, the attractive quality will be accepted even when it is not fulfilling user needs, and when its level increases it will give excitement and satisfaction to the user. In other words, the lowest level of attractive quality is at neutral in terms of the satisfaction–dissatisfaction dimension and it goes up to the excitement level as the needs are fulfilled. It never goes down to the zone of dissatisfaction.

For example, usability, reliability, safety, and minimal functionalities are must-be qualities, whereas novelty, distinctive functionality, and beauty (appeal for Kansei) can be perceived as attractive. For this reason, managers, planners, engineers, and designers tend to focus more on the attractive quality than on the must-be quality. But it should be said that the attractive quality can be effective only when the must-be quality is fulfilled. The attractive quality that lacks consideration of the must-be quality is quasi-attractiveness.

2.4.6   Satisfaction and Significance

As was described earlier, ISO 9241-11:1998 regarded satisfaction as a part of usability. But this is a very narrow view. Users can be satisfied when the product/service has desirable functionality, has a higher level of performance, has good reliability, is safe, and so on. In Figure 2.7, this is represented as a big arrow from objective quality in design to subjective quality in use. Furthermore, users will be satisfied if the product/service is effective and efficient in the real context of use. This is represented as a big arrow from the objective quality in use to the subjective quality in use. In other words, satisfaction is the ultimate dependent variable to all quality characteristics whether they are objective or subjective.

This nature of ultimacy of satisfaction was confirmed by the concept dependency test that was described in Section 2.4.4. And satisfaction, included in the subjective quality in use in Figure 2.7, is the utmost concept to all other quality characteristics.

In ISO 9241-11:1998, satisfaction is defined as "freedom from discomfort, and positive attitudes towards the use of the product." This definition is true but insufficient as the definition of utmost quality characteristics. Satisfaction is somewhat different from other characteristics in subjective quality in use because it is not limited to the emotional reaction and is not limited to any specific aspect of the product/service. It represents the comprehensive acceptance of the product/service with ultimate positive feelings. This is related to the fact that the big arrows (especially of perception) are directed to the subjective quality in use even from the objective quality characteristics.

Satisfaction in Figure 2.7 is attached with the label of significance. The significance is the degree to which the concept of a product/service is meaningful for the goal achievement. Our life is full of goal achievements, although the type of goal differs in relation to the individual lifestyle and other social factors. Hence, some products/services that have much significance to person A may have less significance to person B. Also, there are products/services that are significant to many people, while others are significant only to a limited number of people. The concept of significance can be easily understood if we consider the relationship between the graphical user interface (GUI) and visually impaired users. Although there is such technology as the screen reader, the GUI is fundamentally insignificant for blind users.

Significant products/services will give users satisfaction, whereas insignificant products/services will not. Hence, significance and satisfaction have almost the same meaning; the former refers to the meaning of the product/service and the latter to the resulting emotion inside users.

2.5 FINAL REMARKS

Usability and UX are related with each other but are quite different concepts. Usability is part of objective quality in design, and UX is related to the sum of objective quality in use and subjective quality in use. Distinguishing quality in design and quality in use is valid for clarifying the causal relationship for the product/service. Separating the objective quality characteristics and the subjective quality characteristics is necessary for understanding two different aspects of the product/service.

Furthermore, the evaluation methods for usability are the methods that should be used during the design and are different from the methods for UX that should be used during usage. Evaluation methods for UX are useful to understand the factors influencing the degree of satisfaction in the course of the usage of products/services in real situations by real users. Hence, the result of UX evaluation includes rich information by real users in the real context of use so that the information should be fed back and be used for improving the next version in the planning phase or the designing phase.

3 Users and Artifacts

The goal of user engineering is to improve the quality of life of users in terms of the artifacts that they use. In other words, it is to provide a meaningful artifact to users so that they are able to feel satisfaction.

Our environment is surrounded by and is filled with many artifacts, some of which are visible and some that are invisible. You may have purchased a product recently, but many other artifacts may exist in your life and environment from the past. Some of the artifacts that you are using today may have been handed down by people living decades ago but they are reasonably acceptable in the context of modern everyday life. You consciously or unconsciously accept these artifacts because they have significance in your life. And you are living in this world today with a certain level of satisfaction surrounded by these artifacts.

Take a look around your environment. Most of the artifacts you are using are quite similar to those that your parents and grandparents were using, with the exception of some electronic devices and products made from new materials.

For example, in the kitchen, there are cups and plates, which are items that have been used for many centuries even though the styles have changed according to the tastes of the people at any particular point in time. The same is true of knives, forks, and spoons. The Western style of dining was fixed centuries ago because it was thought to be the ultimate form of the artifact. In Eastern Asia, chopsticks have been used since the ancient times. Today, there are many variations of chopsticks in terms of material, color, and design, but the basic component of the chopsticks are two sticks in everywhere and at any time.

Although people living in Eastern Asia today use knives, forks, and spoons, they use chopsticks for eating Japanese, Chinese, Korean, and Vietnamese foods because such tools are optimized for eating the foods of these particular countries. In this sense, they are significant and give satisfaction to the people using them. Consider a situation where people are asked to eat sushi with a knife and a fork or eat steak with chopsticks. People may be puzzled and frustrated with the disconnect between the eating tool and the food. We can understand how each eating tool is optimized for local food. In other words, eating tools are significant in their own context.

The living rooms of today are furnished with chairs, sofas, and tables that are used in the same way and for the same purposes as in the fourteenth or eighteenth centuries. They provide a certain level of satisfaction to humans, and therefore have significance.

On the other hand, electric and electronic devices have undergone big changes from their early days, and of course, drastic changes from the time when there were no such artifacts. Artifacts have evolved so much today so that people may get greater satisfaction.

The evolution of washing machines is one example. In ancient times (and even in some parts of the world today) when there was no electricity, people washed clothes at the river or the water's edge by trampling, beating, or kneading them. But when electricity was discovered and applied to the chore of washing clothes, the evolution of artifacts started quite rapidly.

The first washing machine had a fan that was activated by a motor. But when people realized that squeezing out wet clothes required physical strength, a washing machine with a squeezer was invented. Because it was sometimes necessary for the handle of the squeezer to be rotated back for loosening the mass of clothes, a washing machine was then invented that had two tubs, one for washing and another for spinning out the water. In this way, the washing machine evolved so that users could be more satisfied. But when it was regarded as troublesome to move wet clothes from the washing tub to the drying tub, a fully automatic washing machine was invented with only one tub that both washed the clothes and then spun out the excess water. This reduced the use of human power and eliminated the wasted time between the end of washing and the start of drying.

The evolution process continued further to the invention of a fully automatic washer-dryer system. The scope of evolution has been enlarged from just washing and will likely reach the stage of complete processing of clothes. In this sense, the next evolutional step for clothes might be the inclusion of automatic ironing. This evolutional history of the washing machine demonstrates that artifacts evolve in the direction of the user becoming more satisfied, and if new products meet users' needs, then they have significance.

On the other hand, many artifacts disappear from the market because they were less significant and less satisfactory to users. One example is a 19-inch TV set combined with an ionizer that was released in 2012 in Japan. The intention of the developer was that viewers could watch TV programs in a comfortable environment. Indeed, ionizers in general are beneficial and the company was good at developing ionizers. But the idea of combining a TV set with an ionizer was not a success, because people who would like to watch TV in a comfortable environment could purchase the ionizer separately. Furthermore, a 19-inch TV set was not attractive to consumers who already had 39- or 42-inch sets. Soon after its release, this product disappeared from the market derivative.

Another example is a digital photo frame that appeared in the market around 2006. While a similar product, digital signage, became popular in public places, digital photo frames for personal use were not widely used. The reason can be attributed to the popularization of cellphones, tablets, and laptops. Users could view photographic images on these multipurpose devices and did not need a single-purpose device in addition to the existing multipurpose products. These negative examples teach us the prerequisites for an article to be accepted.

In conclusion, artifacts evolve until they reach full significance. When they reach this optimal stage, the process of evolution will terminate. At the same time, products with less significance or those that do not match the real need of users will fade away. These are the fundamentals of artifact evolution.

3.1   ARTIFACT EVOLUTION

Every artifact evolves in the direction of satisfying the user in a better way than before. The logic of artifact evolution is shown in Figure 3.1.

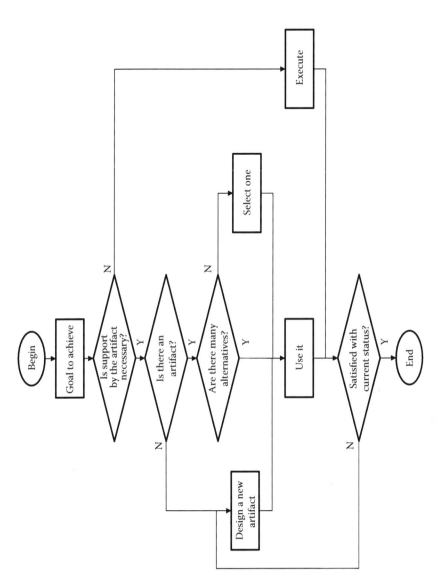

FIGURE 3.1 Flow diagram of the logic of artifact evolution.

First, we have to understand that all artifacts are made to support achieving a goal. Goals may have a layered structure such as goal, subgoal, sub-subgoal, and so on. Hence, the generic goal of "eating something" has many subgoals, such as going to a restaurant, going to a fast-food shop, searching inside the refrigerator, cooking something to eat, and taking something from a natural setting. Going to a restaurant has sub-subgoals such as going to an Italian restaurant, a French restaurant, a Japanese restaurant, a Chinese restaurant, and so on.

For the sub-subgoal of taking something from a natural setting, think about an apple in an apple tree. You will not need any artifact if the apple is within your arm's reach in the apple tree. This is the first junction in Figure 3.1. But you will need an artifact if the apple tree is tall and you cannot reach the fruit; in this case, you know that you will need a ladder. This is the second junction in Figure 3.1. There are, indeed, several ways of getting fruit from a tree. For grapefruit, you need a tree shaker if you're a farmer who has to pick the fruit efficiently. But, in this case, the fruit is not a grapefruit, and furthermore, you will need just a few apples. Therefore, the artifact you need is a ladder. This is the third junction in Figure 3.1. Then you get the apple and are satisfied. This is the fourth junction in Figure 3.1. But if you are in a wheelchair, it is not possible to use a ladder. In that case, you will need another person to get the apple for you. This is humanware (service) and is also an artifact.

If the current artifact is not suitable for goal achievement, it is necessary to design a new artifact, and this step, shown on the left in Figure 3.1, is the path to artifact evolution. The suitability of an existing artifact will be determined in terms of effectiveness and efficiency. Even if many people are satisfied with the current artifact, there will be someone who thinks that it is not effective or not efficient. Before the invention of automobiles, people might have thought that the use of a horse or a horse-driven carriage was the fastest way of reaching your destination. But after the advent of automobiles, people's expectations relating to travel times increased and horse riding became just a hobby.

There are also instances when many people are dissatisfied but there is no adequate artifact. In such cases, collective dissatisfaction will push a new evolution forward. This is seen in the field of medicine: everyone wishes for the advent of wonder drugs against cancer, HIV, and other incurable diseases.

In summary, there are cases where the need for an evolution is overt as well as cases where the need is covert. But in both cases, after the validity of a new artifact is confirmed, people tend to change to the new artifact and discard the previous one.

3.2 ARTIFACT EVOLUTION THEORY

Artifact evolution theory (AET), which was previously called artifact development analysis (ADA) (Kurosu 2008–2009, 2009), is an analytic approach to the optimization of every artifact in terms of a user's goal achievement.

The purpose of AET is to find out what aspects of the current goal achievement are left unsolved or to look for other types of solutions in terms of goal achievement instead of current artifact, in order to promote further evolution. For that purpose, it is necessary to look back at the history of a specific artifact and to survey the diversity of artifacts that are used for a certain goal achievement. In other words, artifacts vary in the two dimensions of time and space. A historical view and a view from cultural anthropology will bring insight regarding artifacts for the same goal achievement, and based on the evidence found in historical change and cultural diversity, we will then be able to consider the possible future direction of the artifact.

3.2.1 Goal Achievement and Artifacts

The relationship between the goal and the artifact is comprised of four patterns, as described in Figure 3.2.

Type SS is the relationship between a single goal and a single artifact. In other words, that artifact is used only for that goal and no other relationship is there. This relationship can rarely be found because most goals will have many artifacts, as in type SM, or most artifacts can be used for many goals, as in type MS. An example of this type of unique relationship can be found between the goal of feeding water to a patient lying in bed and a spout cup as the artifact. No other artifacts can feed water to a patient who is lying down and no other goals can be achieved by the spout cup. An intravenous drip can supply water to the patient, but it cannot eliminate the dryness in the patient's throat. Another example is the relationship between analyzing the constituent of blood and the blood analyzer. No other device can analyze the blood constituent and the blood analyzer cannot be used for any other purpose.

In type SM, many artifacts have been developed for supporting a specific goal achievement. Take the example of listening to music. There are many artifacts provided for this goal achievement, including going to a live concert, listening to a CD or the radio, watching TV, or accessing Internet radio stations or YouTube. Users can select one of

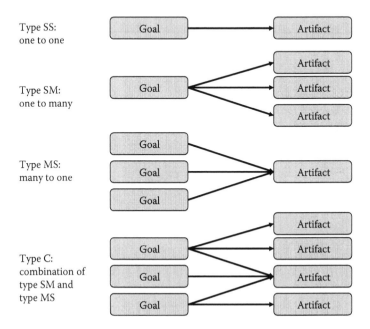

FIGURE 3.2   Relationship between the goal and the artifact.

these considering the nature of their needs (if they want to hear the music right away, if they want it to be of high quality, etc.) and their situation (if they are in their own homes, if they are walking outside or driving, etc.). Another example is a restaurant. In this case, the goal is rather simple, but users will choose the type of restaurant at first depending on the subgoal of the main goal of eating something. Subgoals may include eating something as inexpensive as possible, eating something delicious, eating Asian food, and so on. But when the subgoal is selected, there will be many alternatives that are typical of the restaurant selected. Hence, in this case, the subgoal is what is to be achieved at the specific restaurant selected, and the alternatives on the menu are the artifacts.

Type MS can be found in a multipurpose product such as Kleenex tissue or a multipurpose service such as a concierge. When I surveyed the use of Kleenex, I could find more than 40 goals that it can achieve, such as blowing your nose, cleaning a table or a glass, throwing out chewing gum, and wiping your mouth after eating. Regarding concierge service, there are requests for information about purchasing concert tickets, bathroom locations, nearby convenience stores, transportation from the hotel to the airport, and so on.

Type C is a combination of types SM and MS and the world in general has both these types. For example, instead of asking the concierge for the location of a nearby convenience store, you can ask the doorman or a passerby on the street, or you may walk around and find it for yourself.

3.2.2   PROCESS OF AET

The process of AET will take the following steps:

1. *Survey spatial and temporal variations among artifacts designed for the same goal achievement.* It is necessary to specify whether it is inevitable for an artifact to take a certain form and to specify whether there is a chance of it taking another form. This is based on the notion that a certain form of the artifact used in daily life for a long time must have significance, and thus, has provided satisfaction (or at least has not provided dissatisfaction) to users.
2. *Evaluate the goal-adaptability of each variation.* Each artifact should be evaluated regarding quality in use, user characteristics, and context of use in order to specify what kind of problems, including side effects, may still exist in the specific artifact.
3. *Clarify conditions for the artifact.* Conditions for design of the artifact should be clarified so that goal achievement can be supported adequately with satisfaction.

3.2.3   DIVERSITIES AMONG ARTIFACTS

Spatial diversity described in Figure 3.3 includes variations of the artifact according to culture and geographic environment. Culture has many dimensions, as was described in Figure 1.5. Most typically it is ethnic culture and country culture as found in differences in clothes, food and related tools, houses, and art and music. In addition to ethnic culture and country culture, we can find age group cultures that influence design and

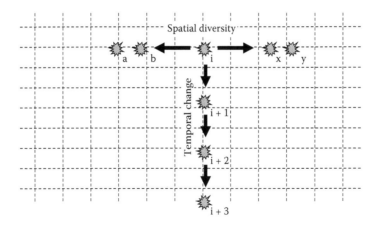

FIGURE 3.3 Spatial diversity and temporal change.

use of tools. For the goal of entertaining themselves, people use various artifacts, such as rattles for babies, building blocks for toddlers, toy musical instruments for young children, and video and computer games for older children, depending on the age group. For the goal of walking, healthy people do not need support artifacts, but elderly people may need walking supports such as canes and walkers or sometimes wheel-chairs. Of course, babies need carriages and strollers before they can walk. We can get this information on the adaptability of artifacts by observing the everyday behavior of people, but more precisely by applying methods of ethnography and folklore.

In addition to culture, environmental factors also make the difference in artifacts so that goal-achieving behavior will adapt to the environment. A computer used in a dusty factory will have to be protected against the dust and will be sealed up; a cellphone used in the desert will have to rely on a satellite instead of an antenna on the ground; a diver's watch will have to be waterproofed, and so on.

Temporal change, described in Figure 3.3, means the change of artifacts within the dimension of time. For analyzing temporal aspects in their full scale, we need the help of archaeology and historical science. These disciplines work to reveal how our ancestors conceived solutions for achieving goals in their everyday lives with similar or slightly different artifacts that we still use today. They also teach us that goals that we as human beings want to achieve are the same over time.

3.2.4 SPATIAL DIVERSITY

As was explained in Section 3.2.3, culture influences the spatial diversity of artifacts. A good example of spatial diversity is an eating tool. Ethnography has revealed that there are three types of eating tools in the world.

The first type of eating tool is fingers, and eating with the fingers is still prevalent in the Middle East, Africa, South Asia, and Oceania. Religious culture is related to eating with the fingers. Among Muslims, eating with the fingers is strictly assigned to the right hand regardless of the dominant hand. The left hand is regarded as unclean because it is used for handling excretion. Some even wash their faces with only the right hand. Eating with the fingers is frequently thought to be primitive, but even Westerners used their hands before Western eating tools became popular. It was also prevalent in Eastern Asia before chopsticks became popular. And even today, Westerners use their fingers for bread and sandwiches and Japanese use their fingers for eating sushi and onigiri (rice balls). Hands and fingers are versatile, and even in areas where eating with the hands is now obsolete, they can be used whenever traditional eating tools are awkward or unsuitable.

The second type of eating tool is Western eating tools, comprised of cutlery including knives, forks, and spoons. An interesting phenomenon is the Southeast Asian way of using Western eating tools. After colonization during the times of European voyages of discovery, Western eating tools came into this area. But because of the manner of cooking in this region, the people of Southeast Asia did not need knives. Food was served in small pieces, and there was no need to cut food on a plate. Hence, people in Southeast Asia now hold their spoons in their right hands and their forks in their left hands.

In addition to regular cutlery, there is a somewhat strange one called the spork (Figure 3.4), which is the combination of a spoon and a fork. Sporks have been used since at least the nineteenth century and some also have the function of a knife. In Japan, they were used for elementary school meals around the 1970s, but they were eliminated because of criticism that they would obstruct the skill acquisition of the traditional eating tool of chopsticks. From the viewpoint of AET, sporks have something of a rationale. They are more economical, safer, more convenient, and weigh less than traditional cutlery. But an eating tool is not just a matter of the tool itself and is closely related to culture and the physical nature of foods. This is the reason sporks are not widely used anymore at least in Japan. Selection of the artifact cannot be achieved by the rationale alone.

The third type of eating tool is chopsticks, which are used in Eastern Asia (Figure 3.5) (Kurosu 2015b). The original region of chopstick use was China, and then expanded to Korea, and finally to Japan. In these countries, knives are used only for cooking and are not used while eating because foods are usually chopped into small pieces. But the use of chopsticks in each of these countries differs a bit.

FIGURE 3.4   Sporks.

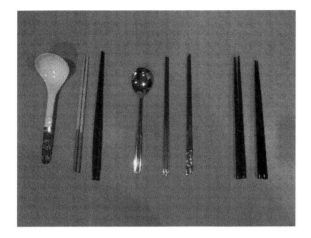

FIGURE 3.5   From left to right: Chopsticks from China, Korea, and Japan.

In China, a Chinese spoon is used in addition to chopsticks for drinking soup. The stick shape can be round or square and the tips are not usually rounded rather than pointed.

In Korea, a flat metal spoon made of aluminum or silver is used in addition to metal chopsticks that have a flat body. It is taboo in Korea to pick up your plate or bowl, so food is picked up with the chopsticks and soup eaten with the flat spoon. For foreigners, using flat metal chopsticks is somewhat difficult, but Korean people usually say that it becomes easier after some training.

In Japan, chopsticks are made of wood, bamboo, or plastic and are usually square-shaped. No spoon is used in a traditional Japanese meal. Instead, people raise up their bowls and drink the soup directly from the bowl.

There is another difference in the use of chopsticks: the direction of the chopsticks on the table. In China and Korea it is normal to place chopsticks vertically, but in Japan they are placed horizontally. The vertical position is not recommended because it looks like the chopsticks are weapons pointing at the person across the table.

But what is written above are formal rules and people actually use eating tools with more flexibility. Today, there are Italian restaurants in Japan that serve spaghetti with chopsticks because the Japanese people, who are accustomed to using chopsticks, find it difficult to eat spaghetti using a fork. Even steak is cut into small pieces before being served (diced steak) and chopsticks are used to eat it.

Regarding the direction of chopsticks on the table or plate, many people place them vertically, as shown in Figure 3.6. It may be because the formal horizontal placement of chopsticks on the table requires three steps to pick them up and is time-consuming:

1. Take the chopsticks by the right hand.
2. Turn the left hand upward and move it below the left part of the chopsticks.
3. Turn the right hand upward and move it below the right part of the chopsticks. Now the chopsticks can be held by the right hand.

FIGURE 3.6   Actual placement of chopsticks in Japan.

However, the vertical placement requires only one step. The different patterns for holding chopsticks are shown in Figure 3.9. In other words, Japanese people are adopting efficiency rather than formalism in everyday situations, as can be seen in Figure 3.6.

Because existing eating tools have long been used in each region, it is difficult to imagine that there will be a big change in the design of eating tools in the future. These tools are designed to adapt to what is being eaten and what is being eaten will not change much, although food choices will increase in this globalizing world. A good example is the popularization of Western foods in East Asia and the popularization of Chinese and Japanese food in the United States and Europe. Many people are using different eating tools depending on what they eat.

3.2.5   TEMPORAL CHANGE

There are two types of temporal change: alteration and coexistence, as shown in Figure 3.7.

In alteration, an artifact that was once popular will be overcome almost completely by a new artifact. The washing machines that were discussed in the first part of this chapter are a typical example of this type. The simple washing machine was replaced by the two-tub washing machine, of which one tub is for washing and another tub is for spinning the water. But when the full automatic washing machine with one tub and two functions of washing and spinning water appeared in the market, two tub washing machine has almost disappeared. And now, at least in Japan, full automatic washer-dryer, that washes, spin the water and dries clothes, is eating away the market of full automatic washing machine. This type of temporal change will occur when a new product includes all the necessary functions of the previous product.

But this strategy does not always succeed. The evolution of the cassette player shows the failure of this type of incremental change. The first generation of cassette

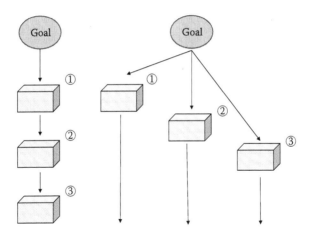

FIGURE 3.7 Two types of temporal change: alteration (left) and coexistence (right).

players only worked with cassette tapes. The second generation was a combination of a radio and a cassette player, and was known as a boom box, or Radi-Casse in Japan. Then the third generation appeared that combined a TV and a boom box. The name of this device was Ra-Te-Casse in Japanese. But the Ra-Te-Casse soon faded from the market, possibly because of the difference in the context of use of the audio-only device and the visual-audio device and because of the small-sized TV screen. Today, Radi-Casses with bigger and clearer sound are sold in stores.

Coexistence is another type of temporal change. In this case, the traditional or previous artifact will still be used in parallel with the new artifact. For example, in the past, live concerts were the only way to enjoy music. But the advent of records, radio, tapes, and other media changed the situation. And although there have been changes in the media used to store music data, and records and tapes were replaced by compact discs (CDs), radio still exists and has even expanded to include Internet radio. Furthermore, live concerts are still being held. This is a typical case of coexistence.

3.2.6 Artifact Evolution and Side Effects

Artifacts evolve in a direction that allow the user to experience a better quality of life and feel satisfaction. Ideally, some of the problems with previous versions will be solved and the total number of problems will decrease. However, this is just an ideal. Sometimes, side effects occur.

Figure 3.8 explains how side effects occur. Assume that the existing artifact or the previous version, shown on the left, has six problems from a to f that should be solved. The designer focuses on the biggest problem, d, from among them and has an idea for solving that problem. Then the new artifact or the new version is designed. But scrutiny on the new artifact reveals that there is a new problem, g, as a side effect of the new design. If the severity of problem g is larger than that of d, the new version will be canceled and will not appear in the market. But if it is not, it will be released to consumers in the market.

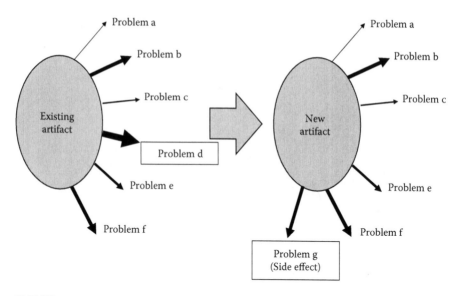

FIGURE 3.8 Artifact evolution and side effect.

For example, traditional books have been a good form of media for getting information. We can read them anywhere there is light and we can start and stop reading at any time. But they can also be voluminous and heavy to carry around. These were the points around which the e-book was designed. We can carry e-books anywhere because they are lightweight and small and we can also read them anywhere and start and stop reading at any time. So all the problems with traditional books seemed to be solved. However, there are side effects associated with e-books, such as they need charged batteries to work, they can't be used in the bath or in places with water, and although sequential access to information is easy, it is not convenient to access information randomly. The battery issue can be solved by loading a long-life battery, the water issue can be solved by a waterproof body, but so far there is no good solution proposed regarding the problem of random access. Hence, we need to find a solution to the navigation issue for the next version.

3.2.7 ARTIFACT EVOLUTION IN TERMS OF FUNCTIONALITY AND USABILITY

We should remember the theory of Jordan that was described in Section 2.4.2 in the context of artifact evolution. Jordan wrote that functionality and usability are not enough and that pleasure is necessary for a product. We could extend his idea to the artifact in general.

As he wrote, functionality is a prerequisite for an artifact. However, engineering-oriented concepts such as performance, reliability, safety, compatibility, cost, and maintainability are also important as the basis of the artifact. Furthermore, usability is also a very important quality characteristic considering that artifacts will be used by users.

Problems described in Figure 3.8 may be related to all possible types of quality characteristics rather than just usability. But it should not be forgotten that usability problems should be included in the list of problems and taken seriously. How deeply usability problems matter can be measured in terms of recognizability, memorability, learnability, operability, user error protection, and accessibility, as was shown in Figure 2.7. Generally speaking, just fulfilling these objective quality characteristics in design will not simply lead to artifact evolution but just to improvement. Drastic artifact evolution should be based on the innovative ideas that are derived from user research in terms of quality in use in a real-world context.

Until the mid-twentieth century, innovative artifact evolution was not accompanied by such user research. Innovative artifact evolution during that time was simply based on intuitive invention by genius. But this fact should not neglect the value of user research. User research will increase the yield rate of innovative ideas and their quality, especially in recent years when almost all goal achievements seems to have been achieved by artifacts and designers are feeling the creative drought. Hence, we should also focus on quality in use, especially in effectiveness and efficiency, of the artifact through observation and interview. Reflection on the evolutionary process of artifacts by AET will surely bring insight for future evolution.

Looking back at the evolutional history of the washing machine that was described earlier, there was a sequence of struggles for the improvement of functionality and usability. But, as Jordan pointed out, pleasure or subjective quality characteristics should be considered in terms of the evolution of artifacts. In a field where functionality and usability are almost saturated, subjective quality characteristics can play a key role in the selection of artifacts. A typical example is the cultural rejection of sporks in Japanese elementary schools. It was not a matter of functionality or usability but the rejection of the artifact based on cultural preference or social allowance.

Hence, we should also consider the subjective aspects, or Kansei aspects, of artifact evolution in addition to the rational or objective aspects.

3.2.8 Artifact Evolution in Terms of Subjective Aspects

The selection of an artifact among different alternatives will be influenced by the factors that were described in Section 1.3. Even for rational or objective aspects such as functionality and usability, subjective factors such as taste, culture, and pleasure influence the selection of a specific alternative. Furthermore, there is an individual difference on the relative weight for functionality, usability, and pleasure. In this section, the influence of culture will be explained in terms of the use of chopsticks in Japan.

When the author surveyed the use of chopsticks, he found that there are different types of norms regarding the good or bad use of chopsticks. Japanese culture has created a set of rules on good manners and bad manners for using chopsticks.

Good manners are DO manners. Good manners are used to appear as smooth and beautiful as the attractive hand motions used in Japanese traditional dance. When picking chopsticks up from the table, it is recommended to take three steps using the right hand first, then the left hand, and finally the right hand (Figure 3.9). The reverse sequence is recommended for placing them back down on the table.

Furthermore, chopsticks placement on the table should be horizontal, unlike in China and Korea where they are placed vertically. In Japan, the vertical direction should be avoided because they are then pointing toward the person sitting across

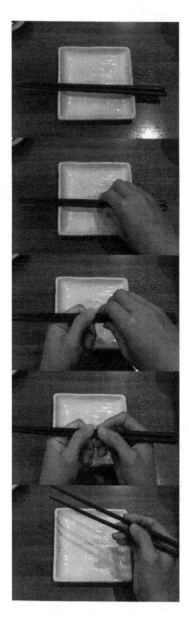

FIGURE 3.9 Taking chopsticks from the table. Left-hand side: good manners, and right-hand side: bad manners (but frequently observed).

the table, which can be interpreted as hostility because the point of the chopsticks might pierce him or her. Of course, it is not meant to physically pierce someone, but it is a visual and mental representation. This rule is positive and is categorized as one of the DOs.

Although it is a cultural constraint, Japanese people often place chopsticks vertically for the purpose of effectiveness and efficiency especially in everyday situations. In other words, the vertical position is more usable than the horizontal position. People also frequently disregard the traditional horizontal position, as was seen in Figure 3.6.

Bad manners are DON'Ts and are categorized as "Kirai-bashi," literally meaning "disliked use of chopsticks." The author collected a total of 43 different Kirai-bashi patterns from many information sources. They include "Arai-bashi," which is washing chopsticks in soup, "Utsuri-bashi," which is moving chopsticks from one plate to another plate without taking food from the first plate, and "Kasane-bashi," which is taking only one specific food without adequately taking food from another plate.

These rules are not for making the manner of using chopsticks beautiful but for making them not look disgraceful or negative.

In conclusion, we can distinguish behaviors in three different layers of positive manners, nonnegative manners, and negative manners, as shown in Figure 3.10. We can also say that, in a more generic sense, the cultural constraints can be classified in these layers in terms of aesthetics. We can easily find other behavioral constraints in such cases as bowing, walking, opening the Japanese room door or "Fusuma," sitting down, and presenting a gift. All these manners or rules are a kind of software in a sense it is a manner of doing things and constitute the part of artifacts.

Further study should be focused on the cultural constraints in foreign cultures. We know that in Korea, plates should not be lifted up while eating food, and that this can be classified as one of the DON'T or nonnegative rules. However, the author does not know yet if there are DO rules in Korea for showing table manners to be beautiful, although we expect to find such rules in the future.

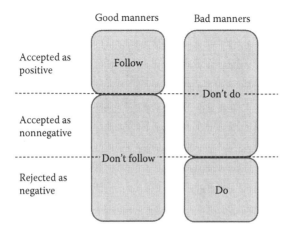

FIGURE 3.10   Manners and acceptance/rejection.

4 Design Process of User Engineering

4.1 BUSINESS PROCESS AND DESIGN PROCESS

In order for users to get satisfaction from the use of artifacts (products and services), the business process, including planning, manufacturing, advertising, selling, and aftercare, should be conducted adequately and deliberately. These processes can be divided into two phases: what to sell and how to sell. Planning and designing belong to the former, and advertising, selling, and aftercare belong to the latter.

4.1.1 BUSINESS PROCESS

Manufacturing artifacts is not just designing them, and collecting and combining materials. There must be a planning stage on what to produce before manufacturing. Of course, designing is important because it will decide the concept, function and usability, and appearance. There must be an evaluation stage afterward to review the manufacturing process and the quality of the artifact. Important information on the evaluation could be obtained from the user or buyer of the artifact after it has been sold in the market. This is the basic scheme of the business process and is known as PDS (plan, do, see) that is said to have been proposed in the early twentieth century by F.W. Taylor.

Sometime later, W.A. Shewhart suggested the PDCA (plan–do–check–act) cycle. W.E. Deming later proposed a revised model, the PDSA (plan–do–study–act) cycle. The critical difference between PDS and these models is the addition of "act," which includes the redesigning activity based on the evaluation (check or study). We can imagine that the "see" in PDS implicitly contains the activity based on the evaluation, but PDCA and PDSA are better because they positioned the act stage explicitly. Act changes the design so that the artifact may fit to the business goal and user needs more accurately.

As I understand the PDCA and PDSA, the check or study stage must be conducted in the real context in order to substantialize the act stage. Usability testing, for example, conducted in the usability laboratory does not reveal real usability problems in the real context of use. In addition, it will not reveal problems of cost, reliability, and other quality characteristics. Hence, the check or study should be conducted in the real context of use.

4.1.2 Design Process

Although these models are for the refinement of the business process as a whole, they can similarly be applied to the design process. This idea can be seen in the process model of ISO 13407:1999 and revised model of ISO 9241-210:2010.

In ISO 13407:1999, the design process starts at "identifying need for human-centered design." The substantial process begins to "understand and specify the context of use," where user research should be conducted to clarify the user and the context of use in order to understand what users are in need of and what problems they may have with existing artifacts, or products and services. Based on findings in the first stage and on existing knowledge of human factors engineering and cognitive psychology, the second stage of "specifying the user and organizational requirements" should be conducted. Requirements as an output of this stage should be solved and implemented in the artifact in the next stage of "producing design solutions." This stage is an iterative process and includes prototyping and evaluation. The evaluation conducted in this stage is the formative evaluation. Next is the fourth stage of "evaluating designs against requirements," in other words, the summative evaluation. If the evaluation results are acceptable, the design process will end because "the system satisfies specified user and organizational requirements." This model was revised as ISO 9241-210:2010 and its content will be explained a bit more in detail in the following sections. But note that these ISO standards are only for the design process and do not cover the whole business process.

Regarding the design process, there is another famous model on design thinking proposed by the d.school at Stanford University (http://dschool.stanford.edu /redesigningtheater/the-design-thinking-process/). In this model, the design process consists of five steps: empathize, define, ideate, prototype, and test. In contrast to ISO 13407:1999 and ISO 9241-210:2010, the first step of this model is to empathize, that is, to deeply understand the current user experience. Understanding the user characteristics and the context of use that is stated in ISO standards is the prerequisite for empathizing the user experience. The second step of defining is similar to the second step of ISO standards. Defining the user's viewpoint is clarifying the user requirements. The third step of ideation to generate the idea for solving the user requirements is not clearly stated in ISO standards, though is implicitly suggested. Generating the idea of a new product or service is quite important and without the idea nothing can be designed. The fourth step of prototyping is the same as producing design solutions but is more clearly expressing the iterative nature of the design. The last step of testing is the same as evaluation in ISO standards. In short, the d.school model is almost the same as ISO standards but is more adequate in describing the ideate process.

Of course, the design process is important so that the artifact may match the requirements based on the user characteristics and the context of use. But the scope of the design process is only within the design, and there is no guarantee that the artifact designed in this step will surely satisfy the user even if the design solution seemed to be an ideal one. In other words, the design is not the only source of success in business. That is to say, what to sell is only the first half of the whole business process and how to sell is the necessary second half.

4.1.3 INNOVATION

Many artifacts are designed and appear in the market. Most business managers aim for the success of their business, thus they look for innovation with the hope of a drastic increase of income. But as Schumpeter ([1926] 1982) and Drucker (1985) pointed out, new technology or new design are not the only ways to achieve innovation.

Schumpeter proposed five types of innovation:

1. New product
2. New production method
3. New market
4. New source of supply
5. New organization

In Schumpeter's scheme, new technology and new design are related only to the first and second items. As an economist, he had a wider view on innovation than an engineer or designer. A new market is necessary for a new artifact to be accepted, a new source of material supply is sometimes needed, and organizational change is necessary to provide a new artifact to the market. In other words, innovation is not only technological change but is change of the organization or the society as a whole.

Drucker also pointed to seven opportunities where innovation can occur:

1. Unexpected occurrences
2. Incongruities
3. Process needs
4. Industry and market change
5. Demographic change
6. Changes in perception
7. New knowledge

In his list, only the last item is related to new technology. Drucker emphasized change of the mind-set of entrepreneurs in terms of issues inside and outside the organization.

In short, innovation may increase the business income, but is caused not only by the new ideas and concepts but also by other business and social issues. The design process is related to innovation but mainly to the improvement of artifacts.

4.2 BUSINESS PROCESS FOR A BETTER USER EXPERIENCE

The business process model integrated with the design process model is shown in Figure 4.1. Because of the nature of service, that is, intangibility, inseparability, heterogeneity, and perishability as Zeithaml et al. (1985) pointed out, products and services will follow a different process.

4.2.1 BUSINESS PROCESS FOR PRODUCTS

The business process starts with planning. The planning should be based on the discussion of what has been found after the user survey that reflects the user experience

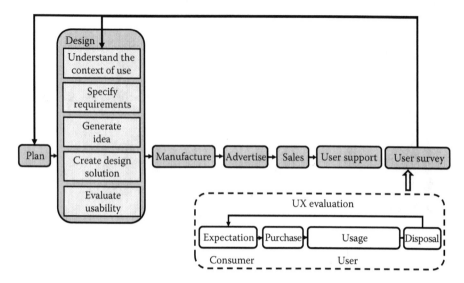

FIGURE 4.1 Artifact lifecycle as a total of business process (gray) and usage process (white).

in the real situation. It decides the product item to be developed and the direction that the item should be manufactured.

Next is the design, whose process is an integration of ISO standards and the d.school model. The first step of the design process is to understand the context of use that is also closely linked to the user survey. The user survey should include all phases of user activity starting from expectation. This phase is done by users as consumers, because they are not yet using the product. The purchase of an artifact is the moment when consumers become users. The next phase—usage—usually lasts for a long time. In this phase, the experience with the product will be stored in memory in three phases: initial, middle, and recent. Usually, users are delighted with the ownership of the product and the initial phase is frequently positive. But it is also the time when novice users will experience difficulty of use and the experience may go negatively in such a case. Such episodes can be stored in memory because initial events in a series of experiences are easily memorized. The middle phase is the time when many things may happen. It may be long depending on the type of product. In the case of food and medicine, the length is usually short. In the case of bags, clothes, and shoes, the length is usually 1 to 2 years. For devices such as home appliances, audiovisual devices, computers, and smartphones, the length is 3 to 10 years. During these long periods, users may experience positive and negative events. But because of the nature of human memory, not all of them are stored in memory, but important experiences are stored whether they are positive or negative. What happened recently is usually stored in memory and remembered easily because of the recency effect. Finally, users will dispose of the artifact (with the exception of foods) because of malfunctioning, low performance, or simply because it is old. All these experiences will be investigated in the UX evaluation during the user survey and the obtained information will be fed back to the design process, especially for understanding the context of use.

The design process goes to the second phase where the requirements are specified. Requirements are not the design itself but the basis for considering the design solution. Sometimes they work as the constraints and sometimes as the basis of idea generation. The next phase in the total design is to generate an idea. This is the phase where individual differences among engineers and designers are great, and something like Osborn's checklist can be used. Based on the idea, the design solution will be created in the next phase, then the usability evaluation will be conducted to the prototype, and the iteration between the refinement of design and the evaluation will improve the degree of perfection.

After the design, the manufacturing will follow. The procedure of manufacturing is different between hardware and software and from product to product. Then the product will be advertised and the information on the product will be delivered to the consumer, thus creating expectation among consumers. When the product is sold in the market, users will take it, try to use it, then decide whether to buy it. This moment corresponds to the time when consumers change into users. After the sale, user support will be given to users.

This is the total lifecycle of the product.

4.2.2 BUSINESS PROCESS FOR SERVICES

In the case of a service, the process is a bit different from products, especially from purchase to usage on the side of users. Usually, the time length of services from purchase to usage is shorter than products due to inseparability and perishability (Zeithaml et al. 1985). Sometimes it is in the order of seconds as in the case of purchasing a ticket. Services at fast food restaurants, hotels, and airports will take more time, but usually in the order of minutes. But user experience with the ticket vending machine or the ATM will be in the order of seconds and minutes. Hence, the difference between the service and the product is not big enough to be differently treated.

There are services that last for more than a month or longer, in cases such as hospital care or house cleaning. These services can be divided into a series of short-term activities, but this kind of activity is similar to dissolve the whole functionality of a device into a set of small units of functions and will have less meaning. The total service at the hospital or by a housemaid should be evaluated as a whole, in a similar manner that products such as computers and smartphone will be evaluated as a whole. In this sense, the business process for services is not completely different from that of products. Although there are such characteristics as the inseparability and the perishability for services, we can treat the user experience for products and services in almost the same manner. Of course, intangibility is quite characteristic to services compared to products. But intangibility may not influence the design process very much.

5 Understanding the User and the Context of Use

Recently, the qualitative approach using interview and observation is becoming popular under the name of business ethnography. This approach is usually conducted in the real environment for real users. Thus, it could be said that some part of the survey using interview, observation, and other methods are covering the user experience (UX) instead of usability. But because the result of surveys will be used for designing the user interface (UI), they can generally be categorized as methods for usability. Remember that the whole process of design concerns only the design process and does not include the whole lifecycle of the product/service including the planning before the design and manufacturing, sales, and actual use in the real context after the design.

5.1 THE USER AND THE CONTEXT OF USE

According to ISO 9241-11:1998, the user can be regarded as part of the context of use. But it would be clearer and easier to say that the user is situated in the context of use and the context of use is surrounding the user. Furthermore, the user and the context of use are independently determined. Of course, there is interaction between them, but they are not always together and each of them should be considered separately. Hence, the user and the context of use are distinguished from each other in this book.

5.2 BASIC APPROACH TO THE USER AND THE CONTEXT OF USE

In order to conduct the survey on the user and the context of use, we will have to consider the following points.

5.2.1 Focus

From the stance of user engineering, the focus of research should aim at improving the quality of life of users and should not primarily aim at increasing the profit of the manufacturer. But users themselves have a limited range of manufacturing power for their own use and they should rely on the power of manufacturers; users can only judge which artifact seems to be the best and decide whether they should buy it. The focus of research will be brought from the business goal, but we should not forget that there is a need of users behind the business goal.

In order to conduct the research on the user and the context of use, we will have to decide the focus of the research. The size of focus could be large if the research aims at general questions such as "the future of automobiles" or be small if it aims at specific questions such as "the future of car navigation system." Regarding general questions, the future of automobiles, for example, should focus on the private car and the public car, the driver and fellow passenger, and various states of the car including running and parking, thus the size of the research becomes larger. But the specific question—the future of car navigation system—will limit the focus only to drivers while running the car (or adjusting the device before running) and those who are already using the system and who are not yet using it.

At the same time when we decide the focus of research, we will have to consider the total length of research. Usually, one interview or observation session should be finished within 2 hours considering the fatigue and the loss of attention on the side of users (informants) as well as the researcher. Research sessions could be repeated a few times, but the total length of sessions should be within 6 to 7 hours. In other words, the more general the focus becomes, the less shallow the research will become. If you would like to ask deep questions, the focus should be limited to the specific questions.

5.2.2 SAMPLING

As was seen in Chapter 1, Section 1.3, there are various dimensions of diversity among users. For understanding the range of each dimension and all the interactions among them, it is necessary to conduct a complete survey. But this is actually impossible for all human beings. Hence, there is a necessity for sampling. Whether the survey approach is quantitative or qualitative, sampling is necessary. Before going further into the sampling method, we will have to think about which approach is better for our understanding of users.

The qualitative approach, typically the questionnaire, has been extensively used in marketing. Usually, it has the advantage that the numerical manipulation and the statistical analysis are possible. Statistical tests give us clear differences, and graphs show the analytic result that is easy to grasp. It gives a macro view on the tendency among users, but does not give a micro view on the mental process of users. Because of the limitation of our cognitive system, we cannot take both of the macro view and the micro view at the same time. If we take the former, we will have to give up the latter and vice versa.

The quantitative approach that gives us a macro view lets us understand the rough tendency of market trends, and it has been useful and is still useful for such a macro approach. But the macro view on the market has not given engineers and designers much information on what users are in need of, how the design of artifact should be, and what users are not aware of for themselves. In order to design the artifact that is really useful for users, it should be designed based on the consideration of such what and how questions. This is the reason why the qualitative approach is necessary in user engineering.

Based on this point, statistical sampling methods that are usually used in the quantitative approach are not used in the qualitative approach. Because of the length of time necessary to conduct the qualitative survey that is usually a week or two, the total number of samples (i.e., informants) we can get in the qualitative approach is limited to 10 to 20. Furthermore, we usually target the whole population (i.e., all the people in the world); the number of samples necessary for the quantitative approach will become a large number, far too large compared to 10 to 20. Even if we limit the population, for example, to male senior people, the sample required is still a big number. In the qualitative survey, we can get only a small number of people and we should take a different sampling paradigm. That is theoretical sampling.

The original idea of theoretical sampling is related to the mental saturation of the researcher. While conducting an interview or an observation, researchers get much information at first. But the amount of new information becomes smaller and smaller as they conduct the qualitative survey with more informants. This is the process of mental saturation and it is the time to stop the survey when there seems to be almost no new information. This is the process of theoretical sampling, and when researchers get saturated they might have established a theory or a hypothesis. The total number of informants in theoretical sampling will vary depending on the size of the research scale. If the scale size is small, the saturation will be fast.

Actually, rigid theoretical sampling is a bit too idealistic. In the ethnographic survey, rigid theoretical sampling is possible and is recommended because researchers have more time, a few months to a year. But the user engineering survey is usually conducted during the business process and the time allowed for the research is limited. Incremental theoretical sampling is usually difficult to follow in such a situation. A mild compromise is group theoretical sampling. In group theoretical sampling, a group of people (8 to 12) will be recruited first and the interview session is reserved for all of them. If the focus is specific and the saturation will be met within that number of informants, the fieldwork will be finished. But if the focus is large and the theory is not yet saturated, another group of people (8 to 12 or fewer) will be recruited. Likewise, group theoretical sampling or stepwise theoretical sampling will be repeated until the theory is saturated.

5.2.3 RECRUITING

In order to collect the necessary number of informants, they will have to be recruited. Some time ago, schoolteachers often used their classroom as an easy way of collecting 30 to 200 samples at once, especially for a questionnaire. But this is not a good method considering classroom samples lead to age bias (i.e., most samples are teens) as well as the ethical aspect.

Mainly there are three types of recruiting: using an informant database, using a recruiting company, and using chain recruiting. Using an informant database is an easy way of collecting samples of specified demographic traits. But it depends on how the database is organized. Also, the same informants could be selected if the size of the database is small. Furthermore, the maintenance of the database, especially the addition of new informants, should be done periodically.

If you have a certain budget, you can ask a recruiting company to specify demographic and other traits. If the recruiting company covers a vast area and various demographic traits, and periodically maintains the database, fulfilling requirement specification is simple and reliable. One thing that should be noted is that the database of a recruiting company is not balanced equally to cover all types of users. Usually, the recruiting company uses the Internet to access the informant. And there is a potential influence regarding ICT (information and communication technology) literacy, income level, and other related factors.

This is the point where chain recruiting will be effective. Chain recruiting starts with friends, relatives, or those who were introduced by the recruiting company. They are the starting points and you will ask them to introduce their friends, relatives, and neighbors by tracing the chain of social networks. Although this method cannot get samples that are almost completely isolated from society, they might be the target of sociological research with special purposes.

Ideally, informants should include a few extreme users who have much knowledge and skills regarding new artifacts; these are the innovators according to Rogers's (1963, 2003) classification. A few users who have little knowledge and experience with new artifacts, called laggards by Rogers, should also be included.

5.2.4 REMUNERATION

In a traditional ethnographic survey, informants are not usually paid for the information they give and the time they spend. This is related to the social position of researcher to the informant. Researchers try to merge into the targeted society, and their information-accessing behavior takes the form of a kind of everyday conversation. They usually do not want informants to take the attitude of trading the information for a reward.

But the fieldwork is a part of business activity and the information obtained from informants will be used to develop a new artifact that will bring the profit in the future. Thus, the proper remuneration should be given to the informants. The amount of remuneration can be estimated by multiplying the hourly income of the informant, the total time required for the session, and the C (coefficient). Hourly income varies based on occupation, but the minimum amount is $10. The total time is not the length of the session but the sum of round trip time for transportation and the length of the session. The C (coefficient) should be 1.5 or higher.

5.2.5 CONSENT FORM

In addition, informants should complete and sign a consent form. The consent form is a document that includes

- Purpose of research
- What informants are requested to do
- Total time of the survey (usually maximum of 2 hours)
- Acceptance of audio/visual recording
- How the data will be used and discarded after the research
- Anonymity of informants in the report or the article
- Right of rejection
- Other notices

The following items should be written at the bottom of the document. Two copies will be made and one is given to the informant.

- Contact information of researcher
- Date
- Signature of researcher and informant

If the theme of research is a critical one, such as a hazardous experience or very private issue, the manuscript of the final report or the article should be sent to the informant. Usually, the fieldwork of user engineering does not go deep into private issues with the exception of artifacts that will be used in a very private behavior, for example, sexual behavior.

5.2.6 INTERVIEW AND OBSERVATION

The qualitative approach consists of the interview and observation. The interview is based on verbal communication with informants, while the observation is based on visual (and sometimes tactile and other sense modalities) information. The interview can be conducted in a laboratory or meeting room. But it is best to conduct it at the informant's place. This is because the real context will make informants relaxed, and it is equipped with relevant artifacts that informants are using. In such an environment, the researcher can note important artifacts, see how various artifacts are organized, and thus understand how the real context is organized. In other words, the researcher not only conducts the interview but, at the same time, should conduct the observation. This point will be explained a bit more in detail. The sections in this chapter mainly focus on the interview or, more precisely, the contextual inquiry embedded in the real context.

The visual record should be done using a camera or video camera, if agreed to by informants. But sometimes, visual information taken by such devices paradoxically does not tell you everything. Usually, there are so many unnamed artifacts and when and how they are used. So, it is recommended to draw a sketch with the name and related information on it in addition to the visual recording. This visual information helps when analyzing the interview data.

5.2.7 RESEARCH QUESTIONS

The research questions are a list of items necessary to clarify the focus of research. This could be a bulleted list of items and may not necessarily be in the form of a question. Usually question items are written in the order of asking them, but as is written in the next section, the actual interview should not strictly follow this order. For an interview of 2 hours, 15 to 20 items are enough. In order to prepare for the case when the scheduled end time comes sooner than expected, important questions should be marked.

5.2.8 Semistructured Interview

Interviews can be classified into three categories: structured, semistructured, and nonstructured. The structured interview is similar to the free-answer questionnaire. The interviewer will ask the questions one by one and the answer will be recorded. Questions will not be skipped nor asked in a different order. This method is not usually adopted in fieldwork but can be used in large-scaled fieldwork using graduate students as interviewers.

The semistructured interview is usually adopted in the user research of user engineering because of its flexibility. Basically, the interview will be conducted following the sequence of research questions. But occasionally the sequence will be changed based on the informant's answers and some new questions will be added. Questions that were marked as fundamental should be asked, but secondary questions can be omitted. What is important in the semistructured interview is not to ask predetermined questions but to reveal the life and mind of users. This flexibility is quite useful for user engineering research.

The nonstructured interview will be conducted without research questions in almost all the cases. What is important to the nonstructured interview is the natural relationship with the informant. Some key information can be obtained from time to time during the time-taking interviews. Because of its naturalistic nature, the nonstructured interview will not end after 2 hours. Sometimes it will be 15 minutes and sometimes 3 or more hours. In short, this is the way that ethnographers are used during their long stay with informants.

In conclusion, for the purpose of user engineering research, the semistructured interview is the best fit.

5.2.9 Contextual Inquiry

As suggested in Section 5.2.6, the interview and the observation should be integrated as the contextual inquiry (Beyer and Holtzblatt 1998). Basically it is an interview, but sometimes informants will use artifacts to help explain what they are doing, and observing (and drawing a sketch or taking the photo or video) such artifacts and their use will help you to understand what the user is doing in their everyday life. For conducting the contextual inquiry, the session should be done at the office or the home of informants so that the actual information on the environment can be obtained. Also, you can get more information than from the interview alone. You can look around and sometimes ask questions about the artifacts in the area.

The contextual inquiry also has the advantage that the research does not interfere with the everyday behavior of informants. Especially when the informants say they do not have much time for an interview, the contextual inquiry based on the observation will give you a set of better information. In the case of observation-based contextual inquiry, you should ask questions by interrupting the work of informants.

5.2.10   SETTING UP THE ENVIRONMENT

If you are to conduct a contextual inquiry, you are to be invited to the informant's environment and you will have almost nothing to do to arrange the environmental setting. In this case, the informant is the host and you are the guest. But, there are cases when the informant will obstinately refuse to have research conducted at their own place, or managers and other stakeholders in the company insist on observing the live session. In such cases, you will have to give up visiting the actual environment and conduct the session in a meeting room. In this case, you are the host and the informant is the guest. So you should arrange the meeting room into a comfortable atmosphere.

It is best to conduct the session using two interviewers. Based on my experience, the combination of a male and a female interviewer created a friendly atmosphere. In the case of two interviewers, the role of each interviewer should be decided in advance. One should be the main interviewer and the other should be the subinterviewer. For an interviewer, there will come the time when reflection on the conversation up to then is necessary. In such a case, the subinterviewer will take the initiative and continue the conversation smoothly.

Observers should not enter the session room so that the informant will not get nervous. Observers can hear the session through a microphone or watch the session through a video camera. The subinterviewer can use a laptop to record the session. The laptop can also be used for communication with observers; there may be the occasion that the observer in another room may have a question that they think should be asked.

The meeting room should be set up comfortably, so there should not be unnecessary objects such as the desktop PC that will not be used in the session. Instead, there should be flowers and plants, and beverages for the informant, in addition to anything that will serve for making a friendly atmosphere.

For the meeting room interview, ask the informant to bring artifacts that are related to the focus of the interview. If you want to ask about a cellphone, the informant should be asked to bring it. If you want to ask about a car navigation system, you should ask the product name and the manufacturer and should prepare the catalog of that model beforehand.

5.2.11   REMOTE INTERVIEW

It can be said that the face-to-face contextual inquiry in the real environment is the best, and the face-to-face interview in the meeting room is second. But there will be times when you would like to (or will have to) conduct user research with those living in a place far away from you. In such cases, the online interview or the remote interview can be conducted using Skype or other audiovisual live communication software. An audio interview that lacks visuals is poor regarding the amount of information but will be acceptable; something is better than nothing.

While conducting the remote interview, especially an audio-only one, much care should be taken regarding the following points:

- You may not be able to feel the physical environment for yourself where the informant is situated, for example, noise level, lighting, temperature, and size of the room.
- Because of the influence of unexpected contextual factors (e.g., the existence of other people) that you may not know, the information you obtain may not simply be compared to that from the face-to-face session.
- Even if a video camera is used, it is difficult to ask the informant to move it to get the image of the artifact you want to see.
- Maintaining the stable mental level is difficult in communication via Internet and a 2-hour interview is a bit much.

5.2.12 CULTURAL ISSUES

Cultural factors, especially ethnic culture including country culture, religious culture, and regional culture, may influence the session. You should check the ethnicity and other cultural factors related to the informants.

Regarding time keeping, some informants are punctual but others are lazy. Regarding the communication, some are talkative and reticent; usually this aspect is more influenced by personality, but there can be cultural differences. Regarding the attitude, some are active and others are passive; this is also related to the personality, but depending on the culture the basic attitude may become aggressively active and, in other cases, silently passive. Some topics or words should be avoided because they may be taboo.

For international online user research, all the aforementioned is true and more care should be taken regarding the language. Not all informants in foreign countries are native speakers of the language that the researcher is using. When the informant seems to have language difficulties, you should pronounce words clearly and slowly. Also, do not use slang or tell jokes; this will puzzle the informants. For these reasons, you should ask a native researcher for support in recruiting informants.

5.2.13 INSTRUCTIONS

Instructions to informants should include almost the same content as the consent form (Section 5.2.5). At the beginning of the interview, the informant should be given the consent form, along with an explanation of each item. Questions are allowed and everything should be clarified before going into the research.

5.2.14 ATTITUDE AND COMPETENCE

Politeness and ethical responsibility are fundamental attitudes that the researcher should take. Furthermore, attentive listening, acceptance of informants' responses, and respect for the initiative of informants are important. In the session, researchers should behave as if they are the apprentices of the informant, who should be regarded as the teacher. But you should not just listen to what the informant says. If there is something that is new to you, you should ask questions about it.

In addition, you should take care that you are a qualified interviewer. There is a set of competence that researchers should have. Kurosu (2016b) and Hashizume et al. (2016) proposed the following list:

1. Planning stage—The ability to understand the business goal, the ability of inference to confirm the validity of the research, and the domain knowledge related to the business goal.
2. Setting stage—The ability of memory to remember the research question, the ability to construct the question item from the research question, and a cooperative attitude.
3. Interview stage—The skill to explain the role of informants, the self-monitoring ability while communicating with the informant, the sensitivity and the knowledge of subjects being talked about, and the multitask ability or the quickness on uptake.
4. Wrap-up stage—The ability to summarize the whole session, the ability of analysis, the ability of remembrance, and an attentive attitude.
5. Analysis stage—The ability of logical thinking and understanding, and the ability of abstraction.

5.2.15 ICE-BREAKING

At the beginning of the session, informants are usually nervous because they do not know the exact questions to be asked and they might worry about giving the correct answers. Because of this, 2 to 5 minutes should be assigned for ice-breaking. Any topic can be used for ice-breaking. Recommended topics are what is related to the informants themselves: their native place, the origin of their name, the fashion they are wearing, and so on. But do not ask questions that have already been asked; some informants get unpleasant when they are asked questions they have already answered. And do not talk about critical issues such as politics, religion, and so forth.

5.2.16 MULTITASKING

The interviewer should conduct dual or more mental information processing at the same time. The interviewer should listen to what the informants are speaking about and understand it. In addition, the interviewer should consider the next question based on what the informant has said, the business goal of the interview, and the research questions. Furthermore, the interviewer should decide which question should be asked before the end of reserved time, and should watch the physical condition and the emotional state of the informant. This multitasking can make the interviewer tired.

5.2.17 RECORDING THE SESSION

If the informants agree to be recorded by visual/audio devices, set the device in an appropriate position. You should not do the recording secretly. That would harm the good relationship with the informant. Devices should be as small as possible, because the smaller the device the less the informant will get nervous.

Although the video camera has the function of audio recording, the sound quality from the distance where it is located is usually poor. This is the reason why the digital recorder should be used as well as the video camera. Of course, it has the purpose of increasing the redundancy of audio recording in the event when the digital recorder does not work well.

You should, of course, check the battery. If it is the charging type, charge the device before the session. And if it is the battery type, insert new batteries before every session.

5.2.18 Wrap Up

The session should be finished at least a few minutes before the reserved time. In any case, you should not go beyond the reserved time. It surely will leave an unpleasant feeling with the informant, and any further requests, for example, to conduct a follow-up interview, might be refused.

At about 10 minutes before the end of the session, review your list of research questions to check if there are any important questions that have not yet been asked. If it is all right, think back over the whole session, remember what was interesting, and ask some additional questions. Just a few minutes before the end of the session, summarize what you heard from the informant for verification. Sometimes, the informant will tell you something they forgot to mention.

At the end of the session, pack the video recorder and documents including the consent form. But it is better to not turn off the digital recorder until you leave the informant's place. Some informants will tell you something meaningful during the exit.

5.2.19 Data Analysis

If you conducted the interview with another interviewer, the debriefing should be done as soon as possible. If you are doing the interview by yourself, you should do the reflection as soon as possible. Even though the audio record is stored in the digital recorder, you should write down your own impressions, what you thought to be important, and construct a primitive hypothesis though it may be fragmental. Human memory is quite fragile and can easily be deformed. So write down your impression in the form of text or in figures while your memory is fresh.

When you go back to your office, methods that will be described in Section 5.3 will be applied for the formal analysis.

5.2.20 Business Goal and Faithfulness

When you write the research report after conducting the analysis that used the methods written in Section 5.3, it is important to remember the business goal and the relationship between the research result and the business goal. Unfortunately, it is frequently observed that the hypothesis you had in mind before the research will be repeated again in the report without any verification by the research data. You should be faithful to the data.

5.3 UNDERSTANDING AND SPECIFYING THE CONTEXT OF USE

Because the amount of original qualitative data obtained from fieldwork is huge and will require much time and effort to be understood by your colleagues and other stakeholders, the data should be compressed and analyzed into a compact and well-organized hypothesis, sometimes called a theory.

The data analysis of the fieldwork is frequently conducted based on intuition. Most of the ethnographies that were written by anthropologists are the result of intuitive analysis. Because subjectivity cannot be completely avoided in the analysis of qualitative data, an intuitive approach must be accepted. But there are many methods that can be used for systematically and objectively analyzing the qualitative data obtained from the fieldwork. These methods include the grounded theory approach (GTA) by Glaser and Strauss (1967), KJ method or affinity diagram by Kawakita (1967), work models of contextual design (Beyer and Holtzblatt 1998), and steps for coding and theorization (SCAT) by Otani (2008, 2011). The abstracted information or the hypothesis will then serve as the basis for organizing the requirement specification.

5.3.1 GROUNDED THEORY APPROACH

The grounded theory approach (GTA) was originally proposed by Glaser and Strauss (1967). It is a methodology to construct a theory (hypothesis) grounded on the field data. Glaser put emphasis on the detailed analysis for small fragmented data, while Strauss emphasized the efficiency by changing the granularity of fragmentation. In Japan, Kinoshita (1999) proposed a modified version called M-GTA.

GTA requires a series of detailed conversion of data to obtain the theory. According to the approach by Strauss, the first step is to fragmentize the original data of inquiries and answers. Then the property and the dimension are extracted and labeled. Following is three levels of coding. The first is the open coding where similar labels are grouped and higher categories are defined. The second is the axial coding where subcategories are related to each other to explain the latent structure of the original data. The third coding is the selective coding where categories are linked. Then the theory is constructed and can be expressed as the category relationship diagram and the story line. These steps are shown in Figure 5.1.

This method is frequently used, especially in nursing science, pedagogy, and other social sciences. Because the theory-making procedure is a bit tough and takes much time, it requires a certain degree of expertise. And even when one has mastered this method, the resulting theory may not be completely the same from analyst to analyst. For this reason, SCAT (Section 5.3.4) is more popular than the original GTA in Japan.

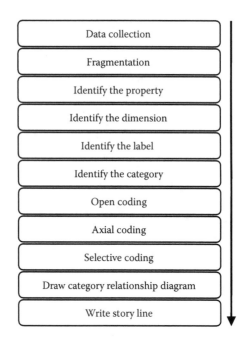

FIGURE 5.1   Steps of the grounded theory approach.

5.3.2 AFFINITY DIAGRAM

The origin of the affinity diagram is the KJ method proposed by Kawakita (1967). Although Kawakita's books were written in Japanese, Kunifuji (2013) wrote an introduction in English. The initial reason Kawakita developed this method was to facilitate the classification of various artifacts and verbal information collected in his anthropological fieldwork.

The principle of this method is quite simple, that is, to let one item move nearer to the other item if they are similar or have something in common or are strongly related. This task will be done on a big table or a big sheet of paper. Usually, each item is written on a small Post-It one item per one sheet of Post-It. Photographs can be used instead of the word or short sentence. Repeat this process until all items are grouped. Care should be taken to equalize the level of similarity or the relationship for all items. In other words, a group for only one item can be made. After grouping all the items, each group will be surrounded by a circle and be given a group name.

In Figure 5.2, an example of the KJ method is shown. In this example, various artifacts used for transportation are classified. The first step is to list all the candidates for classification, which are shown as rectangles. Then similar ones are moved near to each other. Then the first categorization as "Fixed by Rail" or "Fixed by Cable" will be done almost at the same level as other categories. Bus can be uniquely categorized as "Fixed by Road." Next is the second categorization of categories. In this case, "Fixed by Road," "Fixed by Rail," and "Fixed by Cable"

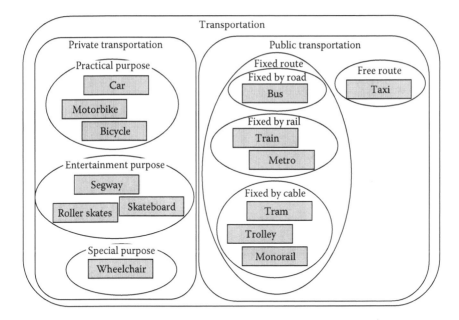

FIGURE 5.2   An example of KJ method, mapping of artifacts used for transportation.

can be categorized as "Fixed Route," which is in contrast with "Free Route" with only one item of taxi. And the third level categorization will be done as "Public Transportation," which is in contrast with "Private Transportation" and finally they are all categorized as "Transportation." Note that the criteria for categorization are different for "Private Transportation" and "Public Transportation," but subcategories such as "Practical Purpose," "Entertainment Purpose," and "Special Purpose" are classified at the same level, and "Fixed Route" and "Free Route" are classified at the same level within each category.

In the affinity diagram, Post-Its are aligned vertically and the group name is written at the top of each row. But considering the procedure after the initial grouping to make a group of groups and draw lines between groups or groups of groups based on the relationship, just making the groups circles seems to be better. The initial group should be located considering the later process of grouping groups and linking groups or groups of groups.

The resulting configuration of groups is something similar to the category relationship diagram of GTA. It is recommended that the initial configuration be broken and separating the items after recording the first configuration and trying to group items in terms of the different criterion for classification.

Lines showing the relationship among circles sometimes mean similarity and other times a causal relationship. Thus, give it some thought when drawing lines between circles of groups. Review the items to see if some of them are a bit different from other items, then re-create groups so that the final configuration will be made.

5.3.3 WORK MODELS

Work models were proposed by Beyer and Holtzblatt (1998). Later, Holtzblatt and her collaborators wrote two books (2004 and 2014) on related subjects. The work model is a part of the contextual design as a total design framework. Contextual inquiry is the first step of the contextual design for obtaining the information from users of which work models will be made for the analysis of information.

Originally, there were five different models: the flow model, sequence model, cultural model, artifact model, and physical model. The flow model represents the spatial relationship among people and artifacts. People are represented as eclipses with the list or roles and behaviors. Artifacts are represented as squares. Eclipses and squares are linked with arrows in terms of their relationship (see Figure 5.3). In contrast to the KJ method that is focused on the collective relationship among artifacts, behavior, and mental activity of people, this model is focused mainly on the social relationships among people who have their own role in the social network. As a result, this model is best suited for analyzing the social network of a small set of people, for example, a project team, family behavior, or the construction of a new shop or restaurant. Drawing this model should be conducted by a group of stakeholders while developing the product or the service. It will help them to understand the social structure among people around the informant and to share the same information about the relational structure.

The sequence model represents the temporal structure of the behavior of informants and related people (Figure 5.4). Originally it was recommended to draw the sequence of what actually happened in a chronological order, but it is better to draw the sequence of events in the normative order in the similar manner as the flow chart that includes the conditional branch. This will help you and your colleagues to understand latent problems that may occur. The flow model and the sequence model thus clarify the space and time structure.

The cultural model is a bit difficult to draw and to use. It is similar to the Venn diagram and shows the inclusive relationship of different cultures that are related to the informant's behavior. Take the example of a construction of a new restaurant. There is the culture (or motivation) of the owner-to-be, culture of the (expected) customer, culture of the manager, culture of the restaurant in the district where it is to be built, culture of the designer (architect), and so forth. Some cultures might be conflicting and some cultures might be dominant (as the source of money). The cultural model will let you understand the kinds of different cultures that are relevant in a single developmental project. And it will let you decide how to compromise possible contradictions.

The artifact model is the model of artifacts used by the user and other stakeholders in the context of work to be done. How the information is organized and by whom is a main concern for making the artifact model. Taking pictures of the artifact is one supplemental way. Photographs copy the visual information in the physical world as it is there but do not tell us what it is, and how and why it was organized and by whom. As a result, it is recommended to draw simple sketches with notes on the paper in addition to the photograph.

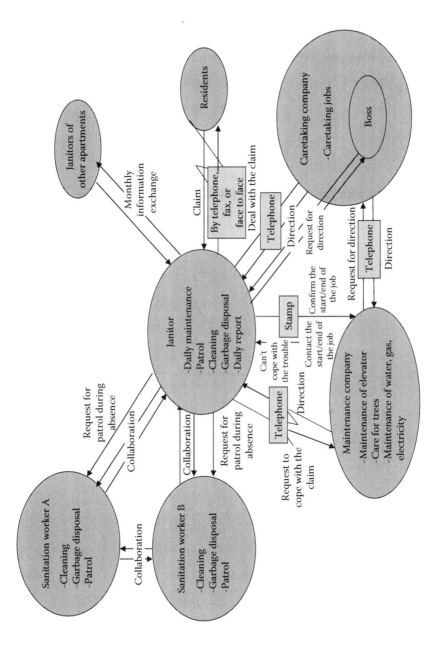

FIGURE 5.3 Example of the flow model (job structure of a janitor of an apartment).

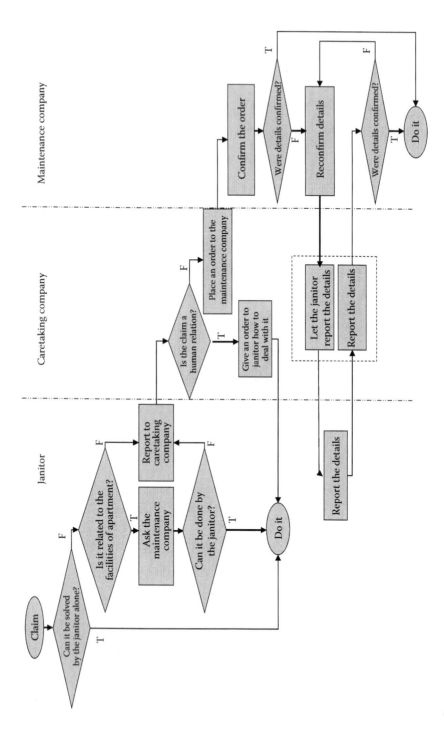

FIGURE 5.4 Example of the sequence model (the case of troubleshooting of a janitor).

Finally, the physical model is a floor layout where the work is done. Sometimes it is important to know where the work is done. But recently, most of our work is not physical but virtual. This model should be used considering the nature of the work.

5.3.4   SCAT

SCAT is an acronym for steps for coding and theorization developed by Otani (2008, 2011). It is similar to GTA in that it is based on the stepwise coding of the transcription of the interview, but is much simpler than GTA and is easier to use. Thus, it is now getting popular in Japan to use SCAT rather than GTA for the analysis of qualitative data.

SCAT uses a Microsoft Excel table. The steps of SCAT start from the transcription of the vocal data of the interview, then each narrative will be separated and be written in the third column of the table. The first column of the table is the sequential number and the second column is the speaker (i.e., interviewer or informant).

An example will be the narrative as follows (this example is translated from Otani 2008):

Interviewer: Then, he became the professor of Chiba University?

Informant: Yes, he resigned from the junior high school. And went to Chiba. Finally, he became an emeritus professor at an industrial university in Kyushu. He was a professional photographer. The university was owned by the Dai-Nippon cellophane company. His home was located in my hometown, and he came back to the town after WWII. He thought he had to do something to earn money and got a job at the junior high school by teaching science. And because my hometown was in a countryside, I was much interested in scientific issues. And that was the starting point of my career. Because the teacher was a professional photographer, I purchased an old-fashioned camera named "BEST" when I was a junior high student. I enjoyed taking pictures and developed them. Because the teacher had a connection with the research laboratory of the film company Fuji Film, they sent us samples of color films before color films appeared at retail stores.

"Notable words and phrases" that attracted the attention of the analyzer will be picked up from the original text and will be written in the fourth column. In this case, it is as follows:

- "Because my hometown was in a countryside, I was much interested in scientific issues."

The next step is to "paraphrase words and phrases" using words and phrases outside of the original text. In this case, they are as follows and will be written in the fifth column.

- "growing environment," "cultural environment," "environmental change caused by an encounter with the teacher"

Then, the "general concept outside the original text that can explain above key phrases" will be summarized and will be written in the sixth column as follows:

- A cultural gap between the existing cultural environment and what the teacher brought about.

Finally, the "theme or the construct considering the context and the whole text" will be written as follows in the seventh column:

- A teacher who can bring a big change to the cultural environment of children.

If there are some points that attracted the attention of the analyzer, they will be written in the eighth column of the table as "Questions and subjects to be discussed further."

These data will be written in the cell of an MS Excel sheet from column one to column eight. After all the text is analyzed, the story line (i.e., the story that can be said at the moment) will be summarized, and the theoretical description and points that need to be discussed further will be written. This is the whole sequence of SCAT analysis.

6 Specifying Requirements

The stage of specifying requirements of the design process summarizes all the relevant information about the artifact to be designed. We already have the following information:

- Plan for the artifact
- Information about the evolutionary process of the artifact
- Information about the actual user
- Information about the real context of use
- General scientific facts on the nature of human beings
- Information about the state-of-the-art technology
- Various constraints

We will have to integrate this information so that we can properly define the requirements.

6.1 PLAN FOR THE ARTIFACT

The decision of the plan for what to design was first brought about from the business process, as was discussed in Chapter 4. Sometimes it is the decision for the improvement of the latest version of the product and other times it is the decision to develop something new. In both cases, the business decision is usually made by the information obtained from market research. Especially when there is an innovative product or service that just appeared on the market from a competing company, the business decision frequently tends to put emphasis on the speed of development rather than the fundamental consideration of the usefulness of the artifact and the degree of user satisfaction.

But from the viewpoint of user engineering, deliberate but quick fieldwork should be conducted before the business decision is made. The fieldwork on the existing related artifact or on the competing new artifact will provide useful information for the business decision of whether you should decide to develop a similar artifact or not jump into the risky development.

When 3D television appeared in the market around 2010, marketing people declared that the new era of television had come, and managers also believed that 3D television was a breakthrough. But the fact was the tide was rapidly on the ebb. The cheap trick of 3D television was the binocular parallax to create depth perception. Although the binocular parallax is a strong psychological cue for depth perception, the 3D visual image created by the binocular parallax contradicts with the physiological visual information made by the accommodation of the lens and the convergence that tell the brain that the visual image is just on a single plane located at a constant distance. This created eyestrain and viewers could not watch the screen for more than 2 hours, because of the fatigue caused by the contradiction between the psychological cues and the physiological cues. Furthermore, engineers noticed that 4K or 16K television was more attractive than the quasi-3D television. This example is quite suggestive that the business decision should be made not based on the desires of marketing people but considering human nature and users' real need.

6.2 INFORMATION ABOUT THE EVOLUTIONARY PROCESS OF THE ARTIFACT

Artifact evolution theory (Chapter 3, Section 3.2) will let us reconsider why current artifact design is as it is and what aspects have not yet been improved. Looking back on the evolutional history of the design will surely give you suggestions as to which direction the artifact should be developed in the near future.

When making the chronological table regarding a specific artifact, you should not forget about the branches that could not grow or stopped being developed. Reconsideration of such underdeveloped and unsuccessful artifacts may sometimes give you a suggestion.

From the viewpoint of artifact evolution theory, there are many cues for innovation. Take for example the vacuum cleaner. Robot cleaners are now becoming popular, but their limitations were also recognized as they began to be used in various situations. The biggest problem is the difference in levels in the house. They cannot be used for cleaning stairways. Another problem is the size of garbage. Users will have to pick up big garbage that the cleaner cannot suction. Spilled liquids is another problem. In order for the robot cleaner to become the main cleaning device in homes, engineers will have to cope with many obstacles.

6.3 INFORMATION ABOUT THE ACTUAL USER

As was written in Chapter 1, Section 1.3, individual differences among users is very vast. Combining all possible alternatives on every attribute will result in almost infinite combinations. A widely used approach to solve this problem is to make a persona description. Persona is a convenient way for directing the mind-set of stakeholders to a certain focus. The information obtained from fieldwork serves as a good basis for describing the persona. In addition, we can learn how personality traits and demographic characteristics are integrated as a whole from the real information about the actual user.

But you should not forget about the negative aspect of persona. Persona is an idealized image of one type of user. Even if several personas are described, they cannot cover the whole range of variety of users; they can only cover the small range of "targeted user" or "intended user." From the standpoint of universal design, this is in line with the user engineering. But all possible users should be considered in terms of the usability of the artifact.

Thus, the best way to use the persona is to regard personas as the center of the possible distribution of diversity. Do not forget about the possibility of usage by other types of user than is described as the persona.

6.4 INFORMATION ABOUT THE REAL CONTEXT OF USE

From the fieldwork, you can also get information about the real context of use. The context of use information includes social, physical, and geographical aspects. Whereas physical and geographical aspects can be perceived by your own senses and can be recorded by photographs and other means, social aspects cannot be seen directly but have a strong effect on the thought and the behavior of an individual.

Social contexts can be found in kinship, occupational, regional, and personal relationships. The context of use of an artifact is situated in the web of such different relationships. Examples include a boy who wears the same shoes as that of his friend, a housewife who purchases a new electric oven that was recommended by her neighbor, a senior person who invests his pension as recommended by the salesperson, and a young girl who folds her clothes in the manner that she was taught by her mother.

If the persona is already described, a scenario can be written to include that information in terms of the use of a targeted artifact by that persona. In the scenario, there is a goal that the persona wants to achieve; the social, physical, and geographical contexts in which the persona is situated; the artifact that the persona selects to achieve the goal; the procedures that the persona took; and the final result with the subjective evaluation by the persona.

In the case of a service, the customer (to-be) journey map can be used. This method allows you to describe the feelings and the thoughts of the persona at each touch point from the start to the end.

But it should be noted that the scenario and the customer journey map are artificial stories and are the outcome of subjective imagination. And the story sometimes reflects unconscious expectation of those who describe them. In other words, they sometimes deviate from the actual context of use. In order to avoid such subjective deviations, it is recommended to insert junctions in the story by which the story goes into different ways. One way of inserting a junction is to split the story into a positive and a negative direction. Adopting other contrasting situations could be useful such as long term versus short term, alone versus with another person, good weather versus bad weather, long distance versus short distance, good taste versus bad taste, and so on.

6.5 GENERAL SCIENTIFIC FACTS ON THE NATURE OF HUMAN BEINGS

Human factors engineering and psychology (especially cognitive psychology) have accumulated much information on the nature of human beings. There are handbooks, textbooks, compendiums, and other sources of information. But such basic knowledge, as listed next, should have been acquired by the designer at the beginning of the project. References should be used only for searching for information that is outside of your common sense boundary.

Human factors engineering
- Typical measures (e.g., regular height of table, optimal distance to the screen)
- Factors affecting a comfortable environment (e.g., temperature, luminance)

- Major factors influencing fatigue
- Classification of human error

Psychological measurement
- Weber's law and Fechner's law
- Rating scale and Likert scale
- Level of scale (i.e., nominal, ordinal, interval, ratio)
- Concept of independent variable and dependent variable

Psychology
- Structure of eye and nature of eye movement
- Concept of figure and ground
- Law of prägnanz of Gestalt psychology
- Factors of depth perception
- Concept of constancy (i.e., size, shape, lightness)
- Color system (e.g., Munsell) and the knowledge of color mixture
- Fundamental model of human information processing
- Norman's seven-stage model of action
- Rasmussen's three-level model of human behavior
- Affordance and signifier
- Concept of metaphor
- Concept of mapping
- Declarative memory and procedural memory
- Episodic memory and semantic memory
- Recognition and recall
- Magic number
- Schema and script
- Concept of mental model
- Learning curve
- Fundamentals of emotion and motivation

Quantitative approach
- Statistical sampling and confidence interval
- Basic concept of statistics (e.g., statistical test, experimental design)
- Basic concept of multivariate analysis (e.g., principal component analysis, clustering)
- Basic skill to use R language

Qualitative approach
- Research questions
- Semistructured interview and contextual inquiry
- Basic requirements for interview research
- Work model, KJ method, and SCAT

Interface designers in a broader sense should know by heart the knowledge of facts and rules listed here.

6.6   INFORMATION ABOUT STATE-OF-THE-ART TECHNOLOGY

Converting an idea into the design requires the knowledge and skill about various tools and technology. There are traditional requirements that conventional designers have to master, such as materials science, mechanical engineering, fundamental dynamics, and printing technology.

But designers today have to master interface technology or the knowledge of human–computer interaction. What is important here is that designers should not necessarily become engineers. It is, of course, better to have the knowledge of signal processing or circuitry design. But if designers don't have it, they can still do their job effectively. What designers should know is what kind of technologies are currently available, what such technologies enable, and the relationship between the input and the output.

For example, knowledge about wearable interface includes information about size, weight, shape, resolution, power consumption, and price of sensors. Designers do not necessarily need to have the skill of programming driver software. Instead, they should know about the variations of wearable devices and have knowledge about what can be used in which situation to realize what kind of goal. It is better if designers have skills about electric circuitry and programming. But those kinds of jobs can be achieved by colleagues who have the skill of engineering.

Knowledge about state-of-the-art technology will, thus, be the basis for building up the requirement by separating what can be done and what cannot yet be done.

6.7 VARIOUS CONSTRAINTS

6.7.1 TIME AND MONEY

Major constraints for the development of an artifact are time and money, of which managers have the power to decide. It is quite unfortunate that we are not living in the dream world where there are no time limits and infinite amount of money. Another constraint may be the lack of person who has the necessary skill to complete the project.

These negative constraints should be considered when summarizing the requirement specifications. Usually the requirement specifications contain many items to be considered. These items should be distinguished by the level of importance. When you look through the items of high importance and find the possibility of prolongation or overspending of the budget, you should negotiate, or sometimes face off, with the manager. It is a minimum requirement to develop a meaningful artifact. Developing a meaningless artifact is a total loss of time and money. Even though there may be the sunk cost effect, a meaningless project should be stopped.

6.7.2 SUSTAINABILITY AND ECOLOGY

Sustainability and ecological aspects are positive constraints that are necessary for our society to be a continuous one. These constraints are important in terms of the conservation of energy and material.

Since the use of shale gas began in the 1990s, the view to the limitation of crude oil seemed to have become a bit positive. Until then, many talked about the future lack of oil, and many alternative energy sources were sought including sunlight, wind, tide, heat of the earth, and nuclear fuel. Such talk seems to have faded. Fossil fuel including shale gas is limited even if its life span has become longer. Someday, the end will come and we will have to restart the same discussion as before.

Most artifacts today use electricity as the energy source and electricity is now heavily dependent on underground resources. Chances are we will find another source for generating electricity even if fossil fuel will be almost consumed in the future. So the energy issue may not be as crucial in terms of the design of artifacts, especially in the twenty-first century.

But the materials issue is more serious. Even though ethylene can be made from shale gas more inexpensively than from crude oil, our civilization today is too dependent on underground materials such as iron, copper, and rare metals as well as gas and oil. Materials for artifacts come mostly from underground. We will have to design artifacts that use the material by recycling existing artifacts. Using renewable resources is our obligation for our descendants. Adopting wood as the material for artifact is one way to solve this problem. But it should be accompanied by the deliberate tree planting and forest restoration activity.

6.8 CONSOLIDATION

By integrating all the aforementioned information, the requirement specification can be summarized. The plan for the artifact (Section 6.1) should be the core of the requirement, the information about the evolutionary process of the artifact (Section 6.2) will support of the direction of the development, information about the actual user and the real context of use (Sections 6.3 and 6.4) will help the realization of the plan, the general scientific facts on the nature of humans (Section 6.5) will improve the feasibility of the plan, and information about the state-of-the-art technology (Section 6.6) will make the plan be up to date. And, as was written in the section on various constraints (Section 6.7), if any difficulty is incidental to the plan, stick to the core importance of the plan again and again or, finally, make the decision to quit the project. Note that, as was described in Chapter 2, an artifact that will give little user satisfaction or that is not significant in the context of use should not be developed.

7 Generating Ideas

When the requirement specification is finalized, the next step is to generate ideas for solving the problem based on the requirement. It is usually said that designers are better in generating new ideas compared to engineers (Winograd 1996). So one way to enrich the idea-generation process is to learn something from the mental process and attitude of designers.

7.1 DESIGN THINKING

Rowe (1987) was the first to use the term *design thinking*. In his context, the term meant "the characteristics of cognitive process among designers," and his intention was to apply this concept to all aspects of design activity including the architecture.

Based on my experience working with designers for 7 years, the characteristics of a designer's mental process is twofold: one is not to stick to the logic in the general sense and another is to be flexible as far as possible.

Illogicality does not mean that designers cannot understand the logic of everyday life. But what I did observe while we were discussing something was that the communication with them gradually or suddenly turns away from the main logic (of which I thought to be the logical path) and they start talking about something different that is far away from the logical path. For me, it was a jump of logic. But for them, it was the next step on the path we were taking together. I experienced these logical jumps so many times. I do not know whether this kind of mental process is the result of training that they had at design school or is the accumulated result of their everyday job of design. Another possibility is that people with this type of mental process become designers. This difference can be termed *divergent thinking*, in contrast to convergent thinking (Guilford 1959). Convergent thinking is the inductive or deductive approach used by engineers and scientists. On the contrary, divergent thinking features fluency, flexibility, originality, and elaboration.

Thus, flexibility can be positioned as a part of divergent thinking, but is quite important in characterizing the mental process of designers. Flexibility is the ability to generate many ideas to solve a problem. This is quite similar to the idea of lateral thinking proposed by De Bono (1967). On the other hand, convergent thinking is similar to the concept of vertical thinking.

In short, what we can learn from the thinking process of designers is divergent thinking and lateral thinking that is flexible and multifocused.

7.2 OSBORN'S CHECKLIST

Characteristics of designers' way of thinking have been rearranged and summarized into guidelines by De Bono, Osborn, and others. Although designers use such ways

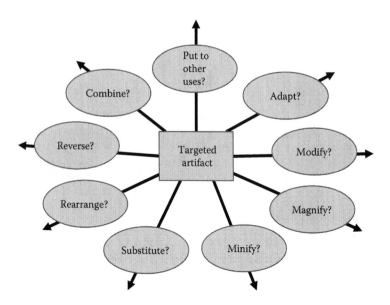

FIGURE 7.1 Nine directions of idea generation from Osborn's checklist.

of thinking without any conscious efforts, we can assimilate their way of thinking by applying such guidelines intentionally. Following is Osborn's (1963) checklist (also see Figure 7.1):

- Put to other uses? As it is? If modified?
- Adapt? Is there anything else like this? What does this tell you? Is the past comparable?
- Modify? Give it a new angle? Alter the color, sound, odor, meaning, motion, and shape?
- Magnify? Can anything be added, time, frequency, height, length, strength? Can it be duplicated, multiplied, or exaggerated?
- Minify? Can anything be taken away? Made smaller? Lowered? Shortened? Lightened? Omitted? Broken up?
- Substitute? Different ingredients used? Other material? Other processes? Other place? Other approach? Other tone of voice? Someone else?
- Rearrange? Swap components? Alter the pattern, sequence, or layout? Change the pace or schedule? Transpose cause and effect?
- Reverse? Opposites? Backwards? Reverse roles? Change shoes? Turn tables? Turn other cheek? Transpose "+/–"?
- Combine? Combine units, purposes, appeals, or ideas? A blend, alloy, or an ensemble?

The key point here is to memorize these guidelines. Make it a habit to think about everything in life divergently.

7.3 POWER OF VISUALIZATION

Drawing the image of something, an object of an interesting shape or moving or changing its shape, a Venn diagram, a graph, or even a word is an externalization of the mental image. Mental image just stored in the working memory in our brain will not grow further, but when it is externalized on paper or is shaped as a simple 3D model, it can be a visual target of your cognitive system. Then you perceive it, and your internal cognitive process can go further. And again, you draw your advanced image on the paper or modify the image that you've already drawn. Then your mental process goes one step further.

Though not yet confirmed psychologically, I think this is the effectiveness of visualization. You can find empirical evidence of this on the desks of many designers. They draw the rough sketch or write a set of words, then crumple the paper, and then take a new piece of paper to draw something anew. They repeat this process again and again until they establish a final first version of the visualized mental model. It is worth imitating what designers are doing in their studio and experience the power of visualization.

7.4 JUMP FROM REALITY

Engineers tend to be bound to various constraints in reality because they are serious and stiff. But sometimes it is worth it to jump into the realm of dreams. Of course, the final design must be adaptive to such constraints. But while generating ideas, you should be out of control of the principles of reality. The world of imagination is full of freedom. And it is sometimes good to spend your time in the principle of freedom.

Because convergent thinking is quite important for designing something in reality, we cannot forget it. But we should not deny the value of divergent thinking. It should depend upon the phase of ideation. For example:

- Early phase—Convergent thinking 10%, divergent thinking 90%
- Middle phase—Convergent thinking 50%, divergent thinking 50%
- Final phase—Convergent thinking 90%, divergent thinking 10%

In this way, the idea of a new artifact will become real at last.

8 Creating the Design Solution

After the generation of idea, it will actually be designed. The traditional design process was straightforward from the upper stage to the lower stage. It is similar to the water stream from the upper reach to the lower reach and will never goes up the stream. This is the reason why traditional design process was called as the waterfall. But today, the design process has been changed to the iterative way because the fundamental deficiency that is detected early in the process can be corrected and the negative influence of the deficiency will be minimized. The most effective iteration is between the design stage and the evaluation stage.

8.1 DESIGNING ITERATIVELY

The earliest design is at the level of the rough sketch, or the rough model for the product or the rough plan for the service. Even though such earliest design has been produced by divergent thinking, the design process hereafter, including the design stage and the evaluation stage, should be conducted mostly in a convergent way.

At each step of iteration, the design should be evaluated. Of course, the evaluation should be conducted in terms of the objective quality in design, especially the usability, and the subjective quality in design, or the attractiveness. Regarding the subjective quality in design, you can make a guess on the satisfaction. But it should be noted that the satisfaction is the quality in use and can never be evaluated correctly at the design stage.

8.2 PROTOTYPING THE PRODUCT

The way of prototyping the product is different for the hardware and the software, although the basic idea is the same. Prototyping the hardware follows the traditional way of prototyping, that is, making a rough mockup using styrene board, polystyrene foam, clay, or wood. Then the blueprint will be drawn by a computer-aided design (CAD) system and the detailed mockup will be made with plastic or other materials. Finally, the functional mockup in which the electric circuit is embedded will be made to confirm the feasibility. Use of a 3D printer is becoming popular and has widened the scope of design. For such new technologies, even the designer should have the sufficient skill of programming and actualizing what is in mind.

Prototyping the software takes the form of a low-fidelity prototype first, then a high-fidelity prototype will be made. But it should be noted that the interactive process is important for the software. Thus, the flowchart and the module diagram should be written before drawing the screen image.

For the low-fidelity prototype, it was recommended to use paper prototyping (Snyder 2003). Snyder recommends a pencil to draw a rough prototype on the paper to make it easier to make modifications. But it takes a long time to draw similar pages with minor differences by hand. Today, there are many software products that can be used for prototyping instead of traditional paper-based prototyping.

8.3 PROTOTYPING THE SERVICE

Service is based on human activity and it cannot be prototyped in the same way as a product. What should be done first is to write the activity scenario where there are agents (e.g., host and guest). The scenario can take the form of the flowchart as the sequence model (see Chapter 5, Section 5.3.3). Then the acting out will be performed. Because the service situation includes various objects in addition to the person, service prototyping needs such objects as well as the agents.

8.4 DESIGN AND EVALUATION

The design process consists of (1) understanding the user and the context of use, (2) specifying requirements, (3) generating ideas, (4) creating the design solution, and (5) evaluation. But the design process that consists of these five steps is just one step in the whole business process that starts from planning, designing, manufacturing, advertising, sales and user support (see Chapter 4, Figure 4.1).

Although human-centered design (HCD) and user-centered design (UCD) focus on the design process, they do not care much about other processes in the total manufacturing process. In other words, they have not committed much to the real usage by real users in the real context of use. From the perspective of usability engineering, only the quality in design is treated so that the usability should be improved. Usability evaluation that was conducted in the usability laboratory is useful and effective for finding out potential problems but does not confirm the quality in use. The actual examination of the quality in use in the real situation, that is, the user experience (UX), was out of scope of the usability engineering.

Then the question arises: When can the UX be evaluated? The answer is that user research after the users have purchased the product is the time when UX is evaluated. In other words, the manufacturing process should not be terminated at the user support, but will and should be continued until the user survey. The reason that the user survey should be conducted after the sales is that it is the responsibility of the manufacturer to confirm if its products and services are positively accepted by users and if there are some negative aspects in their products and services. But how can manufacturers use the information obtained from user research? The answer is shown in Figure 4.1, that is, the information will be fed back to the planning and designing phases.

9 Evaluating Usability

9.1 EVALUATING DESIGNS AGAINST REQUIREMENTS

The evaluation of usability is quite important in the design process to confirm that the product/service does not have any defects in terms of the usability. The difficulty with regard to the discovery of information, memorizing the key information, and learning the procedural steps are some examples of those usability issues. Usability tests and inspection methods are frequently used for evaluating the user interface (UI) of a design. Usability tests are conducted using real users; however, they are usually not held in real or naturalistic environments but in the artificial environment of a usability laboratory. The total test time is usually limited to 2 to 3 hours.

Usually, real problems in the real context of use may be found after long-term use. As a result, it is quite difficult for usability professionals to predict difficulties in the real situation by using only the usability test. The inspection method is also useful for designing the UI because it is quite efficient in detecting usability problems. But the inspection method is usually conducted by the usability professional and is not based on the observation of the behavior of real users. As a result, usability evaluation methods such as the usability test and the inspection method are valuable and significant for improving the UI. But we cannot know how the product/service will be used and accepted by real users in real situations. These usability evaluation methods provide useful information only for designing the UI.

9.2 METHODS FOR USABILITY EVALUATION

Historically, usability evaluation methods required constructing the model of the task to predict the time required for accomplishing the task (e.g., Card et al. 1980, 1983; Kieras 2007). But because more realistic methods could easily bring about more precise predictions, these modeling approaches became an area of academic concern. Psychological experiment was also used to determine the artifact with the best usability among alternatives (e.g., Hirsch 1981). This has been revived as the A/B test (see Section 9.4.4 for more detail). Currently, major usability evaluation methods are the expert review and the usability test.

9.3 INSPECTION METHOD

In inspection methods, usability experts inspect the specification, the mockup, the product, or the manual to detect usability problems based on their intuition. A review of inspection methods has been done by Nielsen and Mack (1994). What made inspection methods famous was the proposal of heuristic evaluation by Nielsen (1993). But it was now recognized that heuristic evaluation should be called expert review.

9.3.1 Heuristic Evaluation

Heuristic evaluation is a simple and easy method to detect usability problems. Usability professionals who have experienced at least 2 to 3 years of usability testing and other usability-related jobs will check mockups or products or documents based on their intuitive judgments. Heuristic evaluation is not recommended for those new to usability engineering because this method is heavily dependent on experience. Furthermore, some engineers dare to apply this method to the product that they designed themselves. This should be strictly avoided because it is quite difficult for engineers or designers to detect usability problems in a design they created.

Nielsen proposed 10 heuristic guidelines (version 2) as follows:

1. Visibility of system status—The system should always keep users informed about what is going on, through appropriate feedback within a reasonable time.
2. Match between system and the real world—The system should speak the user's language, with words, phrases, and concepts familiar to the user, rather than system-oriented terms. Follow real-world conventions, making information appear in a natural and logical order.
3. User control and freedom—Users often choose system functions by mistake and will need a clearly marked "emergency exit" to leave the unwanted state without having to go through an extended dialogue; support undo and redo.
4. Consistency and standards—Users should not have to wonder whether different words, situations, or actions mean the same thing. Follow platform conventions.
5. Error prevention—Even better than good error messages is a careful design that prevents a problem from occurring in the first place. Either eliminate error-prone conditions or check for them and present users with a confirmation option before they commit to the action.
6. Recognition rather than recall—Minimize the user's memory load by making objects, actions, and options visible. The user should not have to remember information from one part of the dialogue to another. Instructions for use of the system should be visible or easily retrievable whenever appropriate.
7. Flexibility and efficiency of use—Accelerators, unseen by the novice user, may often speed up the interaction for the expert user such that the system can cater to both inexperienced and experienced users. Allow users to tailor frequent actions.
8. Aesthetic and minimalist design—Dialogues should not contain information that is irrelevant or rarely needed. Every extra unit of information in a dialogue competes with the relevant units of information and diminishes their relative visibility.
9. Help users recognize, diagnose, and recover from errors—Error messages should be expressed in plain language (no codes), precisely indicate the problem, and constructively suggest a solution.
10. Help and documentation—Even though it is better if the system can be used without documentation, it may be necessary to provide help and documentation. Any such information should be easy to search, focused on the user's task, list concrete steps to be carried out, and not be too large.

Indeed, these are fundamental and very important check items for the user interface design. But there are two problems:

1. Will usability professionals who have already much skill and experience need to be reminded of these items every time they do an inspection?
2. Regarding the usability design guidelines, there are many other guideline items that are related to graphical user interface (GUI) design, accessibility design, human factors engineering, and cognitive psychology. Why is the list limited to 10? Based on this issue, Kurosu et al. (1997) proposed a structured heuristic evaluation method (sHEM) in which all relevant design guideline items are listed but are categorized into several areas. The total inspection session is divided into the same number of subsessions where only each guideline area is the focus. It is based on the psychological consideration that the span of an evaluator's attention is limited.

9.3.2 Expert Review

Because usability professionals do not refer to any guidelines, including the 10 items proposed by Nielsen, while they are trying to detect usability problems, the heuristic method is now generally called the expert review. The fundamental procedure is the same with the heuristic method.

The characteristics of the expert review is that the problem detection can cover the whole aspect of the design but, frequently, not go deep into the interaction process; whereas the usability testing can go deep into the interaction, though it can only check a limited number of possible usability problems. Today, the expert review is frequently combined with usability testing and is conducted prior to the usability testing. Usability professionals can grasp roughly at potential problems with the expert review and make them the basis for making the task scenario for the usability test.

9.4 USABILITY TEST

The usability test is sometimes called the user test, but it is not. We are not testing the user or their ability, but are focusing on the usability of products. We are going to test the usability, not the user. It is deplorable that stakeholders in the field of usability engineering and UX design have a tendency not to be keen to the definition of the concept and the selection of the term.

The history of usability testing goes back to the latter half of the twentieth century when human factors engineers used the method to detect problems related to human factors. When usability engineering arose in the 1980s, the use of usability testing was widened worldwide. Although the inspection method is simple because it does not adopt users in the evaluation, the usability test asks them to come to the usability laboratory or the office where the test will be conducted, and it takes more time and more money than the inspection method. But the advantage that usability professionals and other stakeholders can observe the user's behavior cannot be replaced by other usability evaluation methods.

There are many books about usability testing. Among them is what could be called the standard textbook, *Handbook of Usability Testing: How to Plan, Design, and Conduct Effective Tests*, written by Rubin (2008).

9.4.1 PREPARATION

The usability test is not a naturalistic observation but a laboratory observation. While design activity is in process, artifacts under design may have many usability defects that should be corrected and improved. In the naturalistic situation where users are asked to behave as naturally as possible, it is difficult and inefficient to detect such defects by observing their naturalistic behavior. It is why the laboratory situation or the experimental situation is needed.

Usually, the usability test is conducted in a special room called the usability laboratory, which is usually equipped with a two-way mirror and an observation room from which an observer and other stakeholders of design watch the behavior of the user (Figure 9.1). But if there is no such special room available, the test can be done by using a meeting room. Pay attention to the furnishing of the room so that the user can stay relaxed.

These environments need to be prepared in advance as well as the target device to be tested and the equipment that will help observe and record the user's behavior. There are special devices and software that can record the user's behavior efficiently, but they are not used often.

Recently, remote usability testing is becoming popular especially with website usability. It is an inexpensive way of conducting the usability test for users living in remote places. Usually, the experimenter watches the user's behavior through the screen of the computer that the user is operating and listens to the user's voice

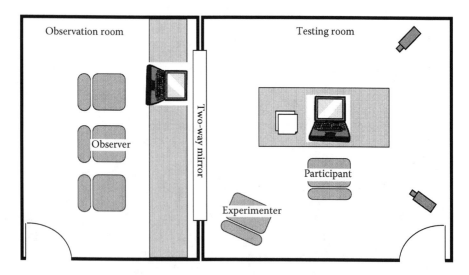

FIGURE 9.1 Typical usability testing laboratory.

through the Internet. It could be a substitution for the face-to-face testing situation, though with the lack of full control of the experimenter.

More important is a task design or the decision on what to ask the user to accomplish. Based on the result of expert review, the crucial problems will be selected and be converted to the task scenario. It should be noted that the total length of the test must not be long. Usually it should be at a maximum of 2 hours. And the number of tasks given to the user should be decided based on the consideration of the execution time of each task. You should print instructions for each task scenario beforehand and hand it to the user after reading it aloud.

Instructions to the user should also be prepared in advance including the request for the thinking-aloud method. Thinking-aloud is a technique developed in the psychology of problem solving for getting information on the experiment participant's thought process. Usually, people do not verbalize their thought process while doing something. But the usability test aims to clarify why users made errors or stopped the operation, and the information about the thought process is necessary. This is the reason why it is necessary to ask users to follow the instruction of the thinking-aloud method. As an exercise for the thinking-aloud method, a simple game such as reversi (also called Othello) can be used.

9.4.2 TESTING

The easiest task will be given to the user first, because many users are still nervous and are afraid of failure. Although they are given the instruction that this is not the test of "your" ability but a test to find defects in the usability of the targeted system, many users still feel that they are being tested. It is important to let them experience success at first.

Then, other tasks will be given considering the time left. During the test, users easily forget the instruction of thinking-aloud, especially when they are absorbed in doing the task. In such cases, a gentle reminder should be given.

When users make an error, it depends on the testing purpose whether to interrupt the user to let them know that their choice was wrong or to let them continue the task until they realize themselves what they are doing wrong. When users cannot proceed with a task and stay silent for a while, you should ask, "What are you thinking now?"

Be sure to end the testing at least 5 minutes before the reserved time. It is quite an unpleasant experience for the user if the testing continues beyond the scheduled time. But if you have reserved much time and you have enough time to discuss the procedure that the user has taken, conduct a retrospective session. In the retrospective session, the video taken during the test will be played and the experimenter will ask more questions about what the user was thinking. It will help to explain the user's behavior more clearly than the simple thinking-aloud method.

It is necessary to write what you observed during the test session just after it has ended. And conducting a debriefing with your colleagues to interpret the user's behavior is quite effective. Sometimes, ideas for solving the usability problem will be generated during the debriefing.

9.4.3 Summarizing the Test

Although it is true that a picture is worth a thousand words, the summary report and the highlight video should be created as soon as possible for the purpose of sharing the information with those who did not attend the testing session. Observers, especially engineers and designers, will get many suggestive ideas while they are watching the testing. But what they remember is not systematic and holistic. The summary report and the highlight video will compensate their lack of professional view.

9.4.4 A/B Test

The A/B test is being used more frequently. The A/B test is a simple experiment and is a derivative of the usability testing that clarifies which alternative of A and B is better. But, what is usually done under the name of A/B test is often far away from the genuine psychological experiment.

First, in the psychological experiment, user groups assigned to A and B should be balanced, especially in terms of the ability for solving the task. But it is difficult in the A/B test situation to balance the subjects in advance. In the usual psychological experiment, the number of users in each group should be decided based on the sampling theory. And the result should be statistically tested. If the null hypothesis is not rejected, you cannot tell whether the values of the measure for A and B are different. Furthermore, you cannot tell the reason why A is better than B. You can get only the number of users that responded in the way that you expected regarding the specific quantitative measure.

10 Evaluating User Experience

10.1 EVALUATION OF USER EXPERIENCE (UX) AND USER RESEARCH

As was described in Figure 4.1 (Chapter 4), the design process includes the stage of understanding the user and the context of use. During this stage, various information on the user and the context of use will be collected in order to specify the design requirements. It is the stage where the user research should be done. Information on the use of the current product and service will surely be quite useful for planning and designing the next version or the next artifact. This is where the information will be used in terms of the real use of artifacts by real users in the real context of use.

It is also true that the process of understanding the context of use in the design process should include the user research in terms of previous products or services or related artifacts. When interpreting the information that was collected during the user research adequately, the past information on the design process of the current version of product should be scrutinized so that stakeholders can understand why the defect by which users are puzzled while using it was thus designed. This information will help them to redesign the product to be more usable.

10.2 METHODS FOR UX EVALUATION

For considering the research methods for the user experience (UX), a temporal model of UX that was proposed in the "User Experience White Paper" (Roto et al. 2011) is useful. It differentiated four UXs as follows.

- Anticipated UX
- Momentary UX
- Episodic UX
- Cumulative UX

From this distinction, we can understand that the temporal aspects of UX are quite important. It should be noted that surveying the UX and evaluating the UX cannot easily be separated, or rather they are almost the same as shown in Figure 4.1. Currently, more than 80 methods are known for surveying/evaluating the UX (All About UX, www.allaboutux.org).

Among them, famous methods include the experience sampling method (ESM) by Larson and Csikszentmihalyi (1983) that asks users what they are doing and what they are feeling by using the cellphone as a questioning tool; the day reconstruction method (DRM) by Kahneman et al. (2004) that is a formalized diary method to ask users to pick a small number of experiences for each day; CORPUS (change oriented analysis of the relationship between product and user) by Von Wilamowitz-Moellendorff et al. (2006) that is a retrospective interview technique using a 10-point rating scale; iScale by Karapanos et al. (2009) that asks the user to draw a graph on usefulness, ease-of-use, and innovativeness using a computer; the UX curve proposed by Kujala et al. (2011) that lets a user draw a curve of experience in terms of attractiveness, ease-of-use, utility, and degree of usage; and AttrakDiff by Hassenzahl et al. (2003) that is a rating scale to rate pragmatic and hedonic aspects.

Methods for the UX evaluation are not yet standardized for usability evaluation partly because of the fact that the concept of UX has not yet been uniquely defined. The methods can be divided into two groups: one is the group of real-time methods (including quasi-real-time methods) and another is the group of memory-based methods. Because the length of time of the UX is generally long, one should select the evaluation method of the UX by the real-time method or the retrospective method. ESM and DRM (including other diary methods) (Bolger et al. 2003) are examples of the former, and CORPUS, iScale, and the UX curve are examples of the latter. AttrakDiff is not a real-time method nor a memory-based method but a rating scale similar to the semantic differential (SD) method and can be used at any time of UX evaluation.

10.2.1   REAL-TIME METHODS

Real-time methods have the advantage of getting the raw impression of users every time the user is asked a question. Because they do not rely on memory, forgetting and distortion will not occur and precise information can be obtained. But it is not easy to use real-time methods for more than 3 to 6 months or even more than 2 weeks because it is difficult for a user to continue waiting for a call or writing experiences in a specified format. In the field of clinical psychology where ESM is also used, it is usually said that 2 weeks is the maximum length for a patient to respond to the call that asks the questions.

10.2.2   MEMORY-BASED METHODS

On the other hand, memory-based methods have the risk of forgetting and distortion of the content that is quite specific to the human memory (Ross 1989; Hsee and Hastie 2006; Norman 2009). But from a standpoint that puts emphasis on the experience at the present time, incidents that are forgotten or distorted might be of small value from the current viewpoint of users, though there may be a chance of suppression. In other words, the exact and correct record of the past may not be necessary for the study of UX. The merit of memory-based methods is that the data gathering can be done in rather a short time and we can collect the abridged information on the user's experience for a long range of time.

In the following sections, we will focus on two methods: time frame diary (TFD) and experience recollection method (ERM) as representatives of the (quasi-)real-time method and the memory-based method. They are methods to get information on how users are/were using the product/service in a certain length of time in the real situation.

10.2.3 Time Frame Diary (TFD)

Time frame diary, or TFD, is a variation of the diary method that asks users to write the details of their daily behavior (Kurosu and Hashizume 2008). Users are asked to fill in the form as shown in Figure 10.1. The form contains 96 time frames each for 15 minutes long for 24 hours. It starts at 00:00 midnight, then 00:15, 00:30, and so on until finally 23:45. Each time frame has two columns: the place column and the behavior column. Informants are given the printed form and are asked to fill in the form every 2 to 3 hours so that the influence of memory can be minimized. Informants are requested to carry the TFD sheet and a pen so that they can record their behavior wherever they go and whatever they do. Instructions will be given to users beforehand so that they focus on the use of the targeted product/service in addition to other daily behavior. Every event relating to the targeted product/service should be recorded correctly and rigidly on the TFD sheet.

Because they are asked to carry the form and the pen with them, the behavior of the user can be recorded in the TFD form almost in real time. Usually this is repeated for 7 consecutive days because the life of most people is repeated week by week. Another reason is that the workload of recording is a bit heavy, and 7 days is thought to be the maximum length.

The interview then follows to specify the feeling and the thought of the user in terms of the targeted product/service. Because the users are shown the TFD sheet that they wrote, the information on the sheet will help them to remember what they thought and felt at each event on the sheet. The example in Figure 10.1 targets the use of a cellphone, that is, when and where the user used it and whether it was a call or an e-mail. The interview should reveal if the user was a sender or a receiver, what is the characteristics of the person to whom the user was communicating, topic of the call or e-mail, feeling of the user while using the cellphone, any problems that occurred during the use of the cellphone, and so forth.

The incentive is quite important for this method because of the load on users. Usually, about $400 is paid for recording seven TFDs and for attending the interview.

10.2.4 Experience Recollection Method (ERM)

The experience recollection method is based on the UX graph (Kurosu 2014b) that is a descendant and a variation of the UX curve (Kujala et al. 2011). In the UX curve, users are asked to draw the curve where the abscissa is the time and the ordinate is the scale value regarding specific aspects of product/service (e.g., the attractiveness, the ease of use, the utility and the degree of usage) with the description of events that occurred during its use from the time when they started using it until the day the survey is conducted.

The author revised this method to describe the event first, then draw a curve, because it is not possible to draw a curve without specifying the location of each event (dots on the graph). This revision is named the UX graph, instead of the UX curve, based on the sequence of the answering procedure. As is described in Figure 10.2, users are first asked to describe the expectation they experienced before the start of use, the experience at the time of start, then during the usage, and at the present time

TFD format   Name ——————— Date ———————— (Saturday) 12 May 2007

Time	Place		What you did	Time	Place		What you did	
0:00	Home	Living room	Drinking	12:00	University	Classroom	Take lesson (English)	
0:15		↓	↓	12:15	Road	Road	Walk (go home)	
0:30		My room	PC (paper work)	12:30	↓	↓	Use cellphone (call)	Ⓢ
0:45				12:45	Home	Living room	Use cellphone (mail)	Ⓡ× 2
1:00				13:00			Have lunch, watch TV	
1:15			↓	13:15				
1:30			PC (mail)	13:30		↓	↓	
1:45		↓	↓	13:45		My room	PC (paperwork)	
2:00		Bedroom	Sleep	14:00	↓	↓	↓	
2:15				14:15	Road	Road	Walk (go to hair salon)	
2:30				14:30	Hairdresser (near my home)		Get a haircut	
2:45				14:45				
3:00				15:00				
3:15				15:15				
3:30				15:30		↓	↓	
3:45				15:45	Road	Road	Walk (go home)	
4:00				16:00	Home	Living room	Use cellphone (mail)	Ⓢ× 2
4:15				16:15		Bedroom	Nap	
4:30				16:30				
4:45				16:45		↓	↓	Ⓡ× 4
5:00				17:00		My room	Use cellphone (mail)	Ⓢ× 4
5:15				17:15	↓	↓	PC (mail)	
5:30				17:30	Road	Road	Walk (go to supermarket)	
5:45				17:45	Supermarket (near my home)		Shopping	
6:00				18:00	↓		↓	
6:15				18:15	Road	Road	Walk (go home)	Ⓡ
6:30				18:30	Home	My room	Use cellphone (call)	Ⓢ
6:45				18:45				
7:00				19:00		↓	↓	
7:15				19:15		Kitchen	Preparation of dinner	
7:30				19:30				
7:45		↓	↓	19:45		↓		
8:00		Living room	Have breakfast	20:00		Living room		
8:15				20:15		↓	↓	
8:30		↓	↓	20:30		My room	Use cellphone (call)	Ⓡ
8:45		My room	PC (mail)	20:45		↓	↓	
9:00		↓	Use cellphone (mail) Ⓢ×2 Ⓡ×2	21:00		Living room	Have dinner	
9:15		Washroom	Outfit	21:15				
9:30		My room	Outfit, PC (mail)	21:30				
9:45	↓	↓	Outfit, PC (mail)	21:45		↓	↓	
10:00	Road	Road	Walk (go to university)	22:00		My room	PC (paperwork)	
10:15	↓			22:15			↓	Ⓡ
10:30	University	Classroom	Take a lesson (English)	22:30			Use cellphone (call)	Ⓢ
10:45				22:45			↓	
11:00				23:00			PC (mail)	
11:15				23:15			PC (paperwork)	
11:30				23:30				
11:45	↓	↓	↓	23:45	↓	↓	↓	

FIGURE 10.1   Example of the TFD for one day.

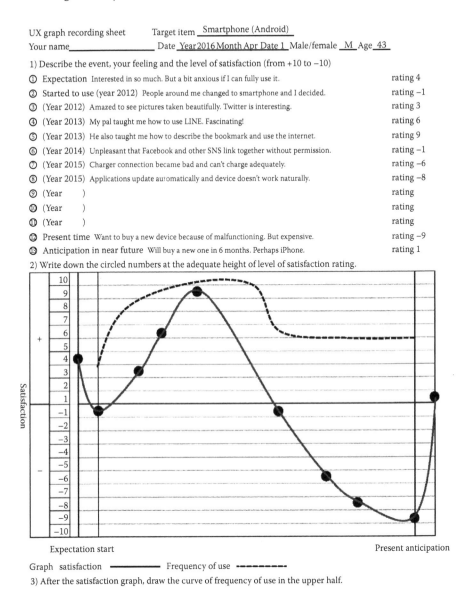

UX graph recording sheet   Target item _Smartphone (Android)_
Your name_____   Date _Year 2016 Month Apr Date 1_ Male/female _M_ Age _43_

1) Describe the event, your feeling and the level of satisfaction (from +10 to −10)

① Expectation Interested in so much. But a bit anxious if I can fully use it.   rating 4
② Started to use (year 2012) People around me changed to smartphone and I decided.   rating −1
③ (Year 2012) Amazed to see pictures taken beautifully. Twitter is interesting.   rating 3
④ (Year 2013) My pal taught me how to use LINE. Fascinating!   rating 6
⑤ (Year 2013) He also taught me how to describe the bookmark and use the internet.   rating 9
⑥ (Year 2014) Unpleasant that Facebook and other SNS link together without permission.   rating −1
⑦ (Year 2015) Charger connection became bad and can't charge adequately.   rating −6
⑧ (Year 2015) Applications update automatically and device doesn't work naturally.   rating −8
⑨ (Year   )   rating
⑩ (Year   )   rating
⑪ (Year   )   rating
⑫ Present time Want to buy a new device because of malfunctioning. But expensive.   rating −9
⑬ Anticipation in near future Will buy a new one in 6 months. Perhaps iPhone.   rating 1

2) Write down the circled numbers at the adequate height of level of satisfaction rating.

Expectation start   Present anticipation

Graph satisfaction —————— Frequency of use − − − − − − − −

3) After the satisfaction graph, draw the curve of frequency of use in the upper half.

FIGURE 10.2 Recording sheet for UX graph.

and in the near future anticipation, with the satisfaction ratings from 10 to −10. In the graph, the ordinate is assigned to the satisfaction as the ultimate measure of UX for the reason described in Section 2.4. The abscissa is extended backward before the start of use so that the expectation or the anticipation can be recorded. The anticipation for the near future is also recorded on the abscissa. Then the graph will be drawn and a dashed curve will be drawn in terms of the frequency of use. After describing graphs, users will be asked questions about their feelings and thoughts.

FIGURE 10.3   Data entry window in Dynamic UX Tool.

In 2016, this method was implemented as the web software "Dynamic UX Graph" (Hashizume et al. Forthcoming). There are five types of episode entries: prior experience, initial experience, experience episodes up to now, current feelings, and future expectations. Data entry windows are shown in Figure 10.3 and the resulting graph is shown in Figure 10.4.

Based on the experiences of conducting this method with more than 300 users, it was realized that the curve goes up and down dynamically depending on the nature of events. In other words, the level of satisfaction as the measure of UX varies so much, hence the magnitude of UX cannot be represented as a single value. Although the value at the present time can be regarded to represent the magnitude of the cumulative UX, the value changes accordingly to the time when the measurement is conducted. Thus, in a week or so, the value may go up or down depending on what kind of episode may happen.

Furthermore, the magnitude of UX in the past is based on the memory. As is well known, memory does not tell us the exact experience. Some episodes may be hidden unintentionally and the magnitude of UX may change based on the evaluation at the present time. This is the point that a memory-based method such as the UX graph is different from a real-time method such as ESM. But we can regard the magnitude of past UXs to represent the evaluation for past episodes from the viewpoint of the present time.

What is important in this method is that we can get the information on the experience as a series of episodes that moves up or down in the graph. And this information will give us useful information on factors that influence the UX. And the important point is that the UX is very dynamic in its nature depending on the episode, the individual characteristics, and the context of use.

Although the UX graph is an adequate method for describing a series of experiences, there was doubt how much the value on the abscissa, or the time of each event, is correct. For example, if two adjacent events A and B were wrongly remembered as A after B, the graph with B after A will give us a different impression compared to the graph with A after B. And the gradient of the graph is not so important. Actually, it was found to be difficult for users to clarify the time when each event actually happened.

Because of this nature of the UX graph, ERM was proposed. In this method, the curve drawing was discarded because of uncertainty of the time of each event. Instead, only the rough time zone is provided to the user. As can be seen in Figure 10.5, "the

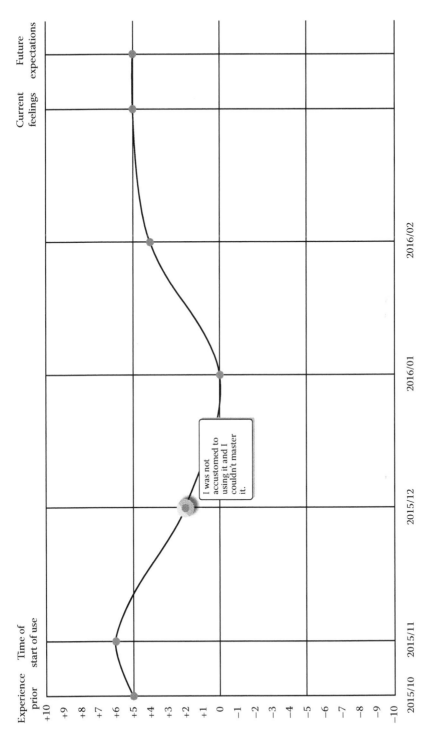

FIGURE 10.4 UX graph drawn on the screen by Dynamic UX Tool.

Recording sheet for ERM: experience recollection method   Target item   Smartphone (iPhone 6)   (Male)/Female   Age   27

1. Write what you experienced at each phase and fill in the evaluation by +10 to −10 rating.

Phase		What you experienced	Evaluation (+10~ −10)
Expectation before purchase		I expected to get the latest model of iPhone on the day of sale.	8
Evaluation at the time of starting to use	Year 2014	I was bewildered for the larger screen compared to my previous model (iPhone5).	5
Evaluation at early days from starting to use		I got used to the large screen soon. And I felt the advantage of the large screen for enjoying the game.	10
Evaluation during the use		The body was bent, but was straighten back by pushing it harder.	5
Recent evaluation		The power loss of battery is unexpectedly fast.	−5
Present evaluation	Year 2016	It's now a must to carry the backup battery.	−5
Evaluation in the near future		I will use this until the next model will appear.	−2

FIGURE 10.5 Recording sheet for ERM.

(Continued)

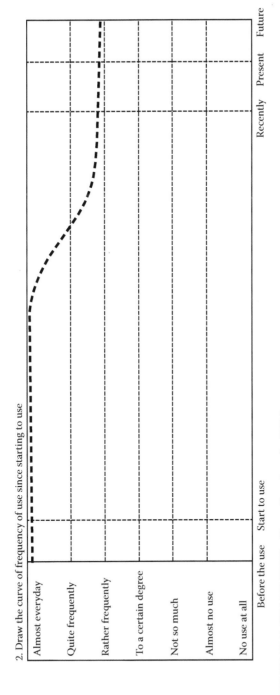

FIGURE 10.5 (CONTINUED) Recording sheet for ERM.

evaluation at early days from starting to use," "the evaluation during the use," and "the recent evaluation" are inserted between "the evaluation at the time of start to use" and "the present evaluation," because such rough time segmentation was thought to be sufficient. Other features of the UX graph were inherited, that is, the satisfaction rating from +10 to −10 and the curve of frequency of usage.

The satisfaction rating is important in the UX graph to realize how much each event recollected from the past experience is evaluated positively or negatively. For the purpose of user research, as described in Figure 4.1, events that resulted in positive feelings should be maintained in the next version of the product or in the future service activity, and events that resulted in negative feelings should be improved in the near future considering the degree of the rating.

In the example described in Figure 10.5, we can see that the big sized screen of the smartphone that astonished the user at the first time was gradually accepted, thus we can think that the size of this smartphone (iPhone 6) can be accepted by (this) user. But the hardware feature or the battery life is a serious issue to be improved in the next version. Fortunately, this user is not leaving this model and is expecting a new, better model to be coming. Because other informants who filled in the ERM format wrote in the same way about the battery life as a big defect (i.e., a large negative aspect), the manufacturer should take this issue very seriously.

The ERM is a qualitative method and it is recommended to confirm the tendency found in the result of ERM by conducting a quantitative method, for example, a questionnaire research with 300 or more informants, based on the hypothesis generated by this method.

10.3   LONG-TERM MONITORING

In ISO 9241-210:2010, it is written that "there is an important difference between short-term evaluation and long-term monitoring" and the follow-up evaluation is recommended "six months to a year after the system is installed." It is true that the short-term evaluation and the long-term monitoring will bring different results. In this sense, the Users Award system that has been held in Sweden (Walldius et al. 2005) deserves applause. In this system, the interview and the questionnaire including 29 criteria in terms of the quality evaluation are conducted 9 months after implementation. But it is not sufficient to conduct one-time evaluation after 6 months to a year.

The monitoring should take place, at least, several times during the whole lifecycle of the product. But it is true that it is difficult for the manufacturer to conduct several surveys for each product it releases. Hence, it is recommended to use a method such as the ERM that uses the memory of users, although there is a small doubt about the reliability due to the nature of human memory.

Roughly speaking, the satisfaction level at the time of purchase or just after the purchase generally tends to be positive partly because of the joy of getting something new, but for those who have less knowledge and skill, it is a time of minor anxiety. And it is certain that such timing is not adequate to measure the degree of UX, though the actual user is using the product in the real context. A few months after the start of using the product, the substantial UX will begin. But the level of satisfaction sometimes may go in diverse directions depending on the ability of users, the match between the product characteristics and the expectation of users, the range of functionality that users discovered and are using, and some unexpected events. The level of satisfaction is not stable and fluctuates from the negative range to the positive range that cannot be represented as a single value. This is the reason why we should look at the UX in the long range of time. In this sense, 6 months to a year after the start of use would be good timing to conduct the ERM survey.

11 Advanced Technology and User Engineering

Technology is progressing very rapidly, and the life of users is also changing accordingly. Thus, it is necessary to try to foresee what might happen in the near future and to consider how to deal with such changes from the perspective of user engineering.

11.1 INTERACTION TECHNOLOGY

The area of interaction technology that is usually called HCI (human–computer interaction) is the most relevant technological area related to the invention and development of new artifacts. But we will have to see how it actually is and how the current situation can be improved.

11.1.1 CURRENT TRENDS IN INTERACTION TECHNOLOGY

The ACM SIGCHI (Association of Computing Machinery Special Interest Group on Computer–Human Interaction) conference is the showcase of new interaction technology. It puts emphasis on the education of HCI, the diffusion of HCI to the world, the linkage with other HCI communities, the support for immature ideas, and other aspects to let HCI technology grow into much a wider field.

But when we look at the content of the conferences, the recent trend is quite different from its early times. When the SIGCHI conference started in the 1980s, there were many challenging categories of presentations, each of which is now a mainstream of their own: virtual reality, computer visualization, multimedia/multimodal interface, experimental approaches using the network, robotics, intelligent interface, interactive art, augmented reality, and mixed reality, in addition to some human-oriented approaches such as human engineering technology, accessibility, psychology, and social issues. These topics were very revolutionary and participants were excited to listen to presentations and join in demonstrations. But now, many presentations tend to apply these seeds into application fields such as medical care, social interface, and industrial products. And other presentations are repeating the dreams of early findings.

Because SIGCHI is the central conference of HCI technology, it could be said that the conference reflects the current general state-of-the-art of HCI fields, but without any surprise or lasting impressions. In contrast such early challenging research such as the *Readings* edited by Baecker et al. (1995), the video "Knowledge Navigator" created by Apple Computer (1987), or the "Aspen Movie Map" and other demonstrations created by the MIT Architecture Machine Group (1978) gave us a real excitement. But today, we do not have such exciting presentations and demonstrations. It is true that the technological development of HCI has changed from that of the 20th century. Although people in academia today are seeking new themes, they have not yet found the novelty in the real sense of the word.

11.1.2   TECHNOLOGICAL APPROACH AND USER ENGINEERING

Here I'll have to point out that technological approaches generally develop new technologies for their own sake. The novelty is what they pursue and they give the highest emphasis on inventing something new that others are not presenting. Engineers and designers tell the happy scenario that their invention will be in favor of increasing entertainment, empowerment, and usefulness in our life. But actually it is just a dream that their invention would like to be. And almost no meaningful verification is done.

Such tendency can also be found in recent product developments. 3D television is one such example that attracted attention at first but soon faded away from the market. Google Glass is not actually used due to many negative reasons. And now the Apple Watch is becoming an obsolete gadget that will not be used by many. Microsoft's Kinect is not still widely used regardless of the seemingly acceptable video that showed its use in the operating room of a hospital. Gesture interface in *Minority Report* is just a scene in a sci-fi movie. Only the technology of virtual reality is actively used in the area of entertainment, although it was advertised in the early years that the practical application in industry would be widened.

From the viewpoint of user engineering, all those failures were inevitable because they did not answer to any real user needs.

In the 1990s when the integration of the humanities and engineering was the focus of interest in academia and industry in Japan, I joined a big project for developing a new system based on the ubiquitous technology. Of course, I was in the humanity group and looked for an application area that the ubiquitous technology could be effectively applied. We conducted fieldwork research in terms of the decision making based on dense communication for a local government service. But when we reached a conclusion, the technology group had already started its approach and designed a device based on some imaginary requirement and imaginary user. The technology group said that a good scenario could now be created to integrate our findings and its design. Oops.

This episode typically describes the traditional tendency of technological development led by engineers. Engineers apt to disregard the real life of users and try to create something that they think to be "good." Of course, engineers are human beings and they are living as users at home or at work. So it seems to be possible that their imagination can lead to a successful invention. Sometimes this approach results in success when their imagination is valid. Until the mid-20th century, this approach made for big successful inventions. But the yield rate of this approach is not high. Furthermore, it is now time that various devices are fulfilling our needs. Today we are coming to the era of big barrier for innovation in its true sense. In such a situation, engineers should change their mind-set and be more sensitive to users' lives by putting more emphasis on understanding users.

11.2 ARTIFICIAL INTELLIGENCE, INTERNET OF THINGS, AND ROBOTS

Artificial intelligence (AI), Internet of Things (IoT; ubiquitous technology), and robots are three hot topics in recent engineering development. Many news programs report about the super progress of these technologies, and many sci-fi movies are made adopting these subjects. Now we are anticipating big changes in our lives with a bit of anxiety that someday human life may be conquered by these technologies.

11.2.1 ARTIFICIAL INTELLIGENCE, INTERNET OF THINGS, AND HUMAN BEINGS

Because AI and IoT are big areas to be discussed from the technological perspective, detailed technological issues will be put aside in this section and we will focus on the relationship between AI and us—human beings.

Precise automatic translation, natural conversation with computers, correct prediction of the weather, speedy information retrieval from a big database, and so on are happy scenarios for us. Furthermore, video monitoring cameras are now almost everywhere in towns and in cars, improving security.

But what would happen if such technological power was handed to a dictator or our society became totalitarian. It is the nightmare described in *Nineteen Eighty-Four* (Orwell 1949). All aspects of our lives would be monitored and checked. There would be no privacy and freedom. Such a society can easily become reality if the social system was to change. And technology is going further without any consideration of the social system where it will be used. The only hope lies in the ethical code among engineers and leaders of our society. There is a similar example on the use of nuclear power that was and is used for negative purposes.

11.2.2 ROBOTS AND HUMAN BEINGS

The growth of robotics is also a recent technological achievement. Whether they are the humanoid or not, robots are already working in our society and everyday life. Currently they are welcoming people at receptions, entertaining people with their lisp talks, working in the production area, watching patients in the hospital, and so on. They can work more than humans do. Robots don't need to sleep, need breaks, or complain about monotonous tasks. Their performance is far better than human beings; for example, they rarely overlook defects. Though currently their use is limited to noncreative jobs, the combination of AI and IoT is threatening to many people. Many humans could lose their jobs. Society could be divided into two groups: those who own and control robots and those who are ousted from the society.

Because the power of robots is not limited to the independent body and they can be connected with one another or to a distributed large-scale computer, each of them can have the power of a super computer. And again the issue of dictators and a totalitarian society should be considered. But even if there are not dictators or society does not become totalitarian, our society will be divided into two groups as mentioned: those who have IT and those who don't.

This is an issue that exceeds the scope of user engineering. But ethical issues and the humanistic view are the fundamentals of user engineering. We will have to keep in mind that users are human, and the happiness of human beings should be considered first. This stance of human-centeredness does not collide with the concept of sustainability and the ecological viewpoint. It is because we, human beings, are not living by ourselves. We should consider the concept of human-centeredness that can harmoniously exist in the optimal environment.

12 Toward the Future

Regarding the development of artifacts that fit users' characteristics and context of use, there are two different prospects on the future of the relationship between artifacts and users: the innovative prospect and the conservative prospect.

12.1 INNOVATIVE PROSPECT

As has been discussed, user engineering pushes forward the development of artifacts that aim to improve the quality of life. The world it describes will be full of significant artifacts and people living in it will be satisfied. In other words, less significant artifacts will be replaced by significant ones, and people will not lose their money and time for unsatisfactory artifacts. People could spend their lives full of effectiveness and efficiency, and have positive feelings throughout the day.

This dreamlike state of things is one goal of user engineering with the prospect of innovation. And all industry stakeholders are aiming toward this goal. This approach was once described by the Japanese Ministry of Economy, Trade and Industry (METI) (2007) as a figure. In the figure, the left-hand side was the improvement cycle and the right-hand side was the innovation cycle. Both cycles start from the accumulation and analysis of data from user research (though it is not clearly expressed in the figure) and the process goes to redesigning of the artifact and the improvement of existing industry in the improvement cycle. On the other hand, the process goes to the clarification of hidden needs, the design of new artifact based on intuition, and the creation of new industry.

Because METI is in charge of the activation of industry, the last step of both processes is summarized to the industrial change. The industrial change is the result of redesigning the existing artifact and creating a new artifact, and that is the key point of this figure. And the source of the information for the redesign and creation is proper user research. Furthermore, METI puts more emphasis on innovation than improvement. When this report was published, UX (user experience) had not yet become a buzzword; instead, Kansei was the focus in Japan. Because of these reasons, we will have to reinterpret this diagram from the viewpoint of user engineering.

Innovation is wonderful if it produces an artifact that is truly innovative and will give benefit to the user. But, as we all know, most "revolutionary" products and services will soon fade away from the market. Can we call such new artifacts that will be forgotten in such a short time "innovation"? Of course not. True innovation should last for a long time and penetrate into every corner of our lives, such as the innovative ideas of the memex (Bush 1945) and mouse (Engelbart and English 1968). These examples, after many revisions, still exist in our lives as the Internet and the current mouse. Thus, true innovations are rare and most of our ordinary efforts should be directed toward redesigning, contrary to the intention of METI.

Sometimes, a genius may be inspired and create a true innovation. But, I would say, it is not the result of the systematic approach in industry. The systematic approach should be more directed toward redesigning. Of course, what is redesigned can have more significance than the previous design. But for successful redesigning, designers should conduct user research in a proper manner and deliberately consider user characteristics and the context of use.

12.2 CONSERVATIVE PROSPECT

The conservative prospect goes against the prevalent opinions among industry people, manufacturers, and designers, as well as marketing people. In short, this prospect is based on a simple discipline of "not to purchase any new artifact" with the condition "until there is a real need among us and the artifact could be a real solution to the problem by which we are suffering."

In Rogers's theory of diffusion of innovations (Rogers 1963, 2003; Rogers and Beal 1958), laggards are treated negatively because they are living in the farthest location from commercialism. But should they be treated negatively? It is true that they do not bring profit to the industry that is redesigning or creating products and services. And it is also true that the stagnation of economic activity will bring bankruptcy to a capitalistic society. But we should also consider the future of human beings in general. There are possible problems, including population explosion, the exhaustion of underground natural resources, and the growing needs in developing countries. We should consider if capitalism could still be the fundamental doctrine of the future. We are now at the turning point from a consuming society to a sustaining society.

Before the advent of the electric iron, the charcoal iron, as shown in Figure 12.1, was used. It was simple, strong, did not malfunction often, and could be used for a long time. Of course, there were such problems as the difficulty of temperature

FIGURE 12.1   A charcoal iron.

control and the necessary procedures before and after the use. It is natural that it was replaced by the electric iron as an innovative artifact. But after the importation of the electric iron, the product lifetime became shorter and redesigned new products fascinated the eyes of consumers.

The history of the iron is suggestive that we are now living in the world of a short consumption cycle where we buy, use for a certain time, and waste. And we are consuming more electricity and natural resources. Will this trend be able to continue forever? I don't think so. Industry will have to make efforts to develop more long-life and energy-saving products so that the user will not waste unnecessary money and the world as a whole can survive for a longer time. This is my hope.

Bibliography

Apple Computer. (1987). "Knowledge Navigator." YouTube video.

Baecker, R.M., Grudin, J., Buxton, W.A.S., and Greenberg, S. (1995). *Readings in Human–Computer Interaction: Toward the Year 2000*, 2nd ed. Morgan Kaufmann.

Beyer, H., and Holtzblatt, K. (1998). *Contextual Design: Defining Customer-Centered Systems*. Morgan Kaufmann.

Bolger, N., Davis, A., and Rafaeli, E. (2003). "Diary Methods: Capturing Life As It Is Lived." *Annual Review of Psychology* 54: 579–616.

Brooke, J. (1996). "SUS: A 'Quick and Dirty' Usability Scale." In *Usability Evaluation in Industry*, edited by P.W. Jordan, B. Thomas, B.A. Weerdmeester, and A.L. McClelland. Taylor & Francis.

Bush, V. (1945). "As We May Think." *The Atlantic Monthly*, July, pp. 36–44.

Card, S.K., Moran, T.P., and Newell, A. (1980). "The Keystroke-Level Model for User Performance Time with Interactive Systems." *Communications of the ACM* 23(7): 396–410.

Card, S.K., Moran, T.P., and Newell, A. (1983). *The Psychology of Human–Computer Interaction*. Erlbaum.

Csikszentmihalyi, M. (1990). *Flow: The Psychology of Optimal Experience*. Harper Perennial Modern Classics.

De Bono, E. (1967). *The Use of Lateral Thinking*. International Center for Creative Thinking.

Drucker, P.F. (1985). *Innovation and Entrepreneurship*. Harper Collins.

Engelbart, D.C., and English, W.K. (1968). "A Research Center for Augmenting Human Intellect." Proceedings of the Fall Joint Computer Conference, San Francisco, California.

Garrett, J.J. (2000). "The Elements of User Experience." http://www.jjg.net/ia/.

Glaser, B., and Strauss, A. (1967). *The Discovery of Grounded Theory: Strategies for Qualitative Research*. Aldine Transaction.

Guilford, J.P. (1959). "Traits of Creativity." In *Creativity and Its Cultivation*, edited by H.H. Anderson. Harper.

Hashizume, A. (2016). "How Can We Get Qualified Interviewers in the Business Scene—In Terms of Education, Experience or Selection?" International Congress of Psychology 2016, Yokohama, Japan.

Hashizume, A., Ueno, Y., Tomida, T., Suzuki, H., and Kurosu, M. (2016). "Development of the Web Tool 'Dynamic UXGraph' for Evaluating the UX." *International Journal of Affective Engineering*.

Hassenzahl, M. (2003). "The Thing and I: Understanding the Relationship between User and Product." In *Funology: From Usability to Enjoyment*, edited by M. Blythe, C. Overbeeke, A.F. Monk, and P.C. Wright, 31–42. Kluwer.

Hassenzahl, M., Bumester, M., and Koller, F. (2003). *AttrakDiff: Ein Fragebogen zur Messung wahrgenommener hedonischer und pragmatischer Qualitat*. Mensch & Computer, Springer.

Hassenzahl, M., and Tractinsky, N. (2006). "User Experience—A Research Agenda." *Behaviour and Information Technology*, pp. 91–97.

Helson, H. (1964). *Adaptation Level Theory*. Harper & Row.

Hirsch, R.S. (1981). "Procedures of the Human Factors Center at San Jose." *IBM System Journal* 20(2): 123–171.

Hofstede, G. (1991). *Cultures and Organization: Software of the Mind*. McGraw-Hill.

Holtzblatt, K., and Beyer, H. (2014). *Contextual Design: Evolved* (Synthesis Lectures on Human-Centered Informatics). Morgan & Claypool.

Holtzblatt, K., and Jones, S. (1995). "Conducting and Analyzing a Contextual Interview." In *Readings in Human–Computer Interaction: Toward the Year 2000*, edited by R.M. Baecker, 241–253. Morgan Kaufmann.

Holtzblatt, K., and Wendell, J.B. (2004). *Rapid Contextual Design: A How-to Guide to Key Techniques for User-Centered Design*. Morgan Kaufmann.

Hsee, C.K., and Hastie, R. (2006). "Decision and Experience: Why Don't We Choose What Makes Us Happy?" *Trends in Cognitive Sciences* 10(1): 31–37.

ISO 9241-11:1998. (1998). "Ergonomic Requirements for Office Work with Visual Display Terminals (VDTs)—Part 11: Guidance on Usability."

ISO 9241-210:2010. (2010). "Ergonomics of Human-System Interaction—Human-Centred Design for Interactive Systems."

ISO 13407:1999. (1999). "Human-Centred Design Processes for Interactive Systems."

ISO 20282-1:2006. (2006). "Ease of Operation of Everyday Products—Part 1: Design Requirements for Context of Use and User Characteristics."

ISO TC159/SC4/WG5. (1988). "WG5 Usability Assurance Sub Group, London Meeting Report."

ISO/IEC 9126-1:2001. (2001). "Software Engineering—Product Quality—Part 1: Quality Model."

ISO/IEC 25010:2011. (2011). "Systems and Software Engineering—Systems and Software Product Quality Requirements and Evaluation (SQuaRE)—System and Software Quality Models."

ISO/PAS 18152:2003. (2003). "Ergonomics of Human-System Interaction—Specification for the Process Assessment of Human-System Issues."

ISO/TR 16982:2002. (2002). "Ergonomics of Human-System Interaction—Usability Methods Supporting Human Centred Design."

ISO/TR 18529:2000. (2000). "Ergonomics of Human-System Interaction—Human Centred Lifecycle Process Descriptions."

Jordan, P.W. (1998). *An Introduction to Usability*. Taylor & Francis.

Jordan, P.W. (2000). *Designing Pleasurable Products: An Introduction to the New Human Factors*. Taylor & Francis.

Kahneman, D., Krueger, A.B., Schkade, D., Schwarz, N., & Stone, A. (2004). "The Day Reconstruction Method (DRM): Instrument Documentation." DRM Documentation.

Kano, N., Sera, N., Takahashi, F., and Tsuji, S. (1984). "Attractive Quality and Must-Be Quality." [In Japanese.] *Hinshitsu* 14(2): 39–48.

Karapanos, E., Zimmerman, J., Forlizzi, J., and Martens, J.-B. (2009). "User Experience over Time: An Initial Framework." *Proceedings of the ACM SIGCHI* 2009, 729–738.

Kawakita, J. (1967). *Idea Generation Method: For Developing the Creativity*. [In Japanese.] Chuo-Kouron-sha.

Kieras, D. (2007). "Model-Based Evaluation." In *The Human–Computer Interaction Handbook*, 2nd ed., edited by J. Jacko and A. Sears. Lawrence Erlbaum Associates.

Kinoshita, Y. (1999). *Grounded Theory Approach*. [In Japanese.] Kobundo.

Kujala, S., Roto, V., Vaananen-Vainio-Mattila, K., Karapanos, E., and Sinnela, A. (2011). "UX Curve: A Method for Evaluating Long-Term User Experience." *Interacting with Computers* 23(5): 473–483.

Kunifuji, S. (2013). "A Japanese Problem Solving Approach: The KJ-Ho Method." In *Proceedings of KICSS 2013*, edited by A.M.J. Skulimowski, 333–338. Progress & Business Publishers.

Kurosu, M. (2005a). "How Cultural Diversity Be Treated in the Interface Design? A Case Study of e-Learning System." HCI International 2005.

Kurosu, M. (2005b). "Human Centered Design—Understanding of User and Evaluation of Usability." [In Japanese.] HQL Seminar.

Kurosu, M. (2006). "New Horizon of User Engineering and HCD." *HCD-Net Journal.*

Kurosu, M., ed. (2008–2009). "Advances in Artifact Development Analysis." [In Japanese.] *Sokendai Journal* 1: 2–3.

Kurosu, M. (2009). "Selection of Means for Learning That Matches to the Characteristics of Learner and the Context of Use—From the Viewpoint of Artifact Development Analysis." [In Japanese.] *Journal of the Open University of Japan* 27.

Kurosu, M. (2010). "A Tentative Model for Kansei Processing—A Projection Model of Kansei Quality." Proceedings of Kansei Engineering & Emotion Research (KEER) 2014 International Conference.

Kurosu, M. (2014a). "Re-Considering the Concept of Usability." Keynote speech at Asia Pacific Conference on Computer Human Interaction (AP CHI) 2014, Bali, Indonesia.

Kurosu, M. (2014b). "UX Curve and UX Graph." [In Japanese.] http://u-site.jp/lecture/ux -curve-and-ux-graph.

Kurosu, M. (2015a). "Usability, Quality in Use and the Model of Quality Characteristics." Proceedings of HCI International 2015.

Kurosu, M. (2015b). "Analysis of Chopsticks from the Viewpoint of Artifact Evolution Theory." [In Japanese.] Proceedings of Japanese Psychological Association 2016.

Kurosu, M. (2016a). "Experience Recollection Method (ERM)." Proceedings of HCI International 2016.

Kurosu, M. (2016b). "Classification of Competences for Interviewers Working in Industry." Proceedings of International Congress of Psychology 2016.

Kurosu, M., and Hashizume, A. (2008). "TFD (Time Frame Diary)—A New Diary Method for Obtaining Ethnographic Information." Proceedings of Asia Pacific Conference on Computer Human Interaction (AP CHI) 2008.

Kurosu, M., and Hashizume, A. (2014). "Concept of Satisfaction." Proceedings of Kansei Engineering & Emotion Research (KEER) 2014 International Conference.

Kurosu, M., Matsuura, S., and Sugizaki, M. (1997). "Categorical Inspection Method—Structured Heuristic Evaluation (sHEM)." *IEEE International Conference on Systems, Man, and Cybernetics*, 2613–2618.

Larson, R., and Csikszentmihalyi, M. (1983). "The Experience Sampling Method." In *Naturalistic Approaches to Studying Social Interaction*, edited by H.T. Reis, 41–56. Jossey-Bass.

Lidwell, W., Holden, K., and Butler, J. (1997). "The Principles of Universal Design," Version 2.0. North Carolina State University, The Center for Universal Design.

Mace, R. (2016). "About UD." *The Center for Universal Design*, https://www.ncsu.edu/ncsu /design/cud/about_ud/about_ud.htm.

Ministry of Economy, Trade and Industry (METI). (2007). "Report on the Initiative for Creation of Kansei Value." [In Japanese.]

MIT Architecture Machine Group. (1978). "Aspen Movie Map." YouTube video.

Moore, G.A. (2014). *Crossing the Chasm*, 3rd ed. Harper Collins. First published 1991.

Morville, P. (2004). "User Experience Design." http://semanticstudios.com/user_experience _design/.

Nielsen, J. (1993). *Usability Engineering.* Academic Press.

Nielsen, J., and Mack, R.L., eds. (1994). *Usability Inspection Methods.* Wiley.

Norman, D.A. (1988). *The Psychology of Everyday Things.* Basic Books.

Norman, D.A. (2004). *Emotional Design: Why We Love (or Hate) Everyday Things.* Basic Books.

Norman, D.A. (2009). Memory Is More Important than Actuality. *Interactions* 16(2): 24–26.

Norman, D.A., and Merholz, P. (2007). "Peter in Conversation with Don Norman About UX & Innovation." http://adaptivepath.org/ideas/e000862/.

Orwell, G. (1949). *Nineteen Eighty-Four.* Secker & Warburg.

Osborn, A.F. (1963). *Applied Imagination: Principles and Procedures of Creative Problem Solving*, 3rd rev. ed. Charles Scribner's Sons. First published 1953.

Otani, T. (2008). "'SCAT' A Qualitative Data Analysis Method by Four-Step Coding: Easy Startable and Small Scale Data-Applicable Process of Theorization." *Bulletin of the Graduate School of Education, Nagoya University* 54(2): 27–44.

Otani, T. (2011). "SCAT: Steps for Coding and Theorization—Qualitative Data Analysis Method with Explicit Procedure, Easy to Set About, and Suitable for Small Scale Data." [In Japanese.] *Kansei Engineering* 10(3): 155–160.

Rasmussen, J. (1986). *Information Processing and Human-Machine Interaction: An Approach to Cognitive Engineering.* Elsevier Science.

Revang, M. (2007). "The User Experience Wheel." http://userexperienceproject.blogspot .jp/2007/04/user-experience-wheel.html.

Rogers, E.M. (1963). "What Are Innovators Like?" *Theory into Practice* 2(5): 252–256.

Rogers, E.M. (2003). *Diffusion of Innovations*, 5th ed. Free Press. First published 1962.

Rogers, E.M., and Beal, G.M. (1958). "The Importance of Personal Influence in the Adoption of Technological Changes." *Social Forces* 36(4): 329–335.

Ross, M. (1989). "Relation of Implicit Theories to the Construction of Personal Histories." *Psychological Review* 96(2): 341–357.

Roto, V., Law, E., Vermeeren, A., and Hoonhout, J., eds. (2011). "User Experience White Paper." http://www.allaboutux.org/files/UX-WhitePaper.pdf.

Rowe, P. (1987). *Design Thinking.* MIT Press.

Rubin, J. (2008). *Handbook of Usability Testing: How to Plan, Design, and Conduct Effective Tests*, 2nd ed. Wiley. First published 1994.

Schumpeter, J.A. (1926) 1982. *Theorie der Wirtshaftlichen Entwicklung 2* [The Theory of Economic Development]. Transaction Publishers.

Shackel, B., and Richardson, S.J., eds. (1991). *Human Factors for Informatics Usability.* Cambridge University Press.

Snyder, C. (2003). *Paper Prototyping: The Fast and Easy Way to Design and Refine User Interfaces.* Elsevier.

Von Wilamowitz-Moellendorff, M., Hassenzahl, M., and Platz, A. (2006). "Dynamics of User Experience: How the Perceived Quality of Mobile Phones Changes Over Time." *In User Experience—Towards a Unified View*, 74–78.

Walldius, Å., Sundblad, Y., and Borning, A. (2005). "A First Analysis of the Users Award Programme from a Value Sensitive Design Perspective." Proceedings of the 4th Decennial Conference on Critical Computing: Between Sense and Sensibility.

Winograd, T. (1996). *Bringing Design to Software.* ACM Press.

World Health Organization (WHO). (2001). "International Classification of Functioning, Disability and Health (ICF)."

Zeithaml, V.A., Parasuraman, A., and Berry, L.L. (1985). "Problems and Strategies in Services Marketing." *Journal of Marketing* 49: 33–46.

Index